Ruthenium - Materials Properties, Device Characterizations, and Advanced Applications

Edited by Yao-Feng Chang

Published in London, United Kingdom

Ruthenium - Materials Properties, Device Characterizations, and Advanced Applications
http://dx.doi.org/10.5772/intechopen.104132
Edited by Yao-Feng Chang

Contributors
Ichiro Yamashtia, Huanwen Han, Kazuyuki Nobusawa, Fumie Takei, Ting-Chieh Chu, Noriyasu Hashida, Adebare Nurudeen Adewunmi, Khotseng Lindiwe Eudora, Ntalane Sello Seroka, Su Huaneng, Thomas Ernst Ernst Müller, Mulisa Maumela, Ndzondolelo Bingwa, Marius Orlowski, Mohammad Al-Mamun, Amrita Chakraborty, Duduzile Zamavezi Nkomo, Maje Phasha, Hein Moller

First published in London, United Kingdom, 2023 by IntechOpen
IntechOpen is the global imprint of INTECHOPEN LIMITED, registered in England and Wales, registration number: 11086078, 5 Princes Gate Court, London, SW7 2QJ, United Kingdom

British Library Cataloguing-in-Publication Data
A catalogue record for this book is available from the British Library

Additional hard and PDF copies can be obtained from orders@intechopen.com

Ruthenium - Materials Properties, Device Characterizations, and Advanced Applications
Edited by Yao-Feng Chang
p. cm.
Print ISBN 978-1-80356-986-4
Online ISBN 978-1-80356-987-1
eBook (PDF) ISBN 978-1-80356-988-8

For EU product safety concerns:
IN TECH d.o.o., Prolaz Marije Krucifikse Kozulić 3, 51000 Rijeka, Croatia,
info@intechopen.com or visit our website at intechopen.com.

Meet the editor

Dr. Yao-Feng Chang is a TD Q&R Memory Reliability staff at Intel. He received a Ph.D. in Electrical and Computer Engineering from the University of Texas at Austin, USA, in 2015. His primary research focuses on emerging electronics and memory devices for advanced technology nodes for storage, computation, and energy-efficient integrated systems. He has published more than 100 journal publications and conference proceedings, and 5 filed patents.

Contents

Preface

Ruthenium (Ru) is a promising next-generation interconnect material for back-end-of-line (BEOL)-compatible processes owing to its low resistance to oxide, high melting point, and low bulk resistivity.

The potential of material properties and device characterizations can be exploited in the immediate future if more efficient hardware and devices are developed that meet the special requirements of high-performance computing in scaling schemes. In this book, we discuss the basic materials properties and potential device characterizations of Ru in various advanced applications, such as non-volatile memory technology (not restrained to any type of device) in combination with various hardware applications. This book provides a platform for interdisciplinary research into materials science with emerging physical substrates. It includes studies in areas such as material and physic modeling, materials physics and analytics, devices in miniature scale, advanced functional circuits, high-speed computing systems and integration for logic applications, and other novel emerging device concepts and circuit schemes for potential biomimetic smart systems.

Yao-Feng Chang
Intel Corporation,
United States of America

Section 1

Materials Properties

Chapter 1

Ru Complex Ion Induces Anomalous Enhancement of Electrochemical Charge Transfer

Huanwen Han, Kazuyuki Nobusawa, Fumie Takei, Ting-Chieh Chu, Noriyasu Hashida and Ichiro Yamashita

Abstract

Electrochemical impedance spectroscopy (EIS) is a highly sensitive observation technique to detect the state of electrode surfaces in solution. A small amount of $[Ru(bpy)_2DPPZ]^{2+}$, a well-known DNA intercalator and fluorescent light switch, has been found to abnormally increase the charge transfer of the mediator $[Fe(CN)_6]^{3-/4-}$ at the surface of carbon electrodes. When a very small amount of the Ru complex is added to the EIS solution, a large impedance decrease occurs. This phenomenon is caused by the carbon electrode, the mediator $[Fe(CN)_6]^{3-/4-}$ and $[Ru(bpy)_2DPPZ]^{2+}$. No other agents are necessary. By adding $[Fe(CN)_6]^{3-/4-}$ and a very small amount of $[Ru(bpy)_2DPPZ]^{2+}$ to the PCR solution, EIS measurements using a PVA-coated carbon electrode could monitor PCR progress in real-time as an increase in impedance.

Keywords: electrochemical impedance spectroscopy (EIS), $[Ru(bpy)_2DPPZ]^{2+}$, hexacyanoferrate, $[Fe(CN)_6]^{3-/4-}$, PCR, carbon electrode

1. Introduction

We are investigating electrochemical biosensors. Electrochemical biosensors have attracted researchers' attention because of their ability to operate under physiological solution conditions and biochemical high salt concentration solutions, simple configuration, and easy handling. In the electrochemical sensor, three electrodes, the working electrode (WE), the counter electrode (CE), and the reference electrode (RE), are immersed in the solution to be measured and detect changes in the charge transfer between redox mediators added to the measurement solution and the working electrode. A highly sensitive sensor can be realized by detecting changes in impedance between WE and CE. Hexacyanoferrate, $[Fe(CN)_6]^{3-/4-}$, of a few mM is often used as a redox mediator. The design of sensor systems varies widely, and to date, a variety of electrochemical biosensors have been proposed and studied [1–9].

We have been applying this electrochemical biosensor for real-time detection of polymerase chain reaction (PCR). PCR is a technique that amplifies the dsDNA of a target by doubling and used for detection of infectious viruses [10, 11], identification of viruses and bacteria [12, 13], diagnosis of tumors [14], gene expression analysis [15] and single nucleotide polymorphisms [16]. It is usual that real-time PCR detection has been realized using a fluorescent dye and optics-based sensing method. However, conventional PCR detection requires a delicate and bulky optical fluorescence measurement system, and heavy equipment, which make the PCR device fragile, not portable, and expensive. Replacing the optics with a small electrical circuit should make PCR robust, compact, and cost-effective. It can be concluded that the use of an electrochemical sensor would be quite advantageous. Although there have been reports of electrochemical real-time PCR detection (quantitative PCR, qPCR) using voltammetry measurement techniques and EIS measurement methods [17–24], we have devised a completely new PCR detection system based on the simple EIS method, but introducing the idea of adding a small amount of a second mediator.

2. Electrochemical impedance spectroscopy (EIS)

We applied electrochemical impedance spectroscopy (EIS) as the biosensor detection technique. Three electrode system was employed, a working electrode (WE) a counter electrode (CE), and a reference electrode (RE), which are immersed into the solution, and the impedance between the WE and a CE is measured by applying AC voltage of about 10 mV with sweeping frequency. **Figure 1(a)** [25–32].

When the current due to the redox reaction is measured, Nyquist plot of the complex impedance is a combination of a semicircle and a straight line with a 45-degree slope (**Figure 1(b)**). This spectrum can be interpreted by Randall's equivalent circuit (**Figure 1(a)**), which consists of the charge transfer resistance of the electrode surface (Rct), the Warburg impedance due to mediator diffusion in the low-frequency range (Rw), the electric double layer capacitance of the electrode surface (Cdl) and the solution resistance (Rs). The semicircle is produced from the parallel junction of Rct and Cdl, and its diameter is approximately equal to Rct. Depending on the solution state, the charge transfer between the mediator and the electrode changes, resulting in a change in Rct which is the sensor output. The straight line with a 45-degree slope in the low-frequency region is the Warburg impedance.

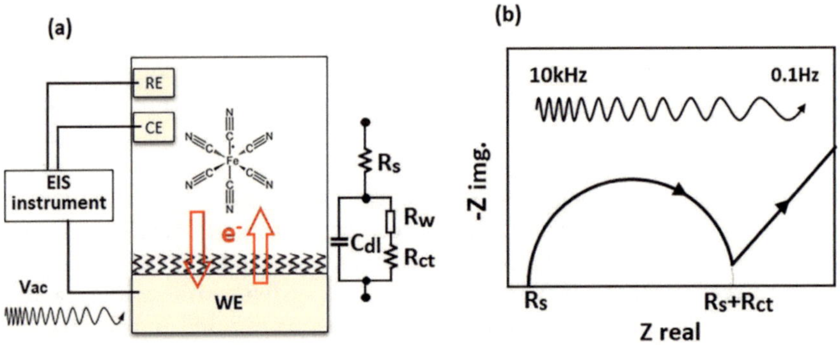

Figure 1.
(a) Schematic drawing of charge transfer between hexacyanoferrate, $[Fe(CN)_6]^{3-/4-}$ mediator, and working electrode (WE) with Randall equivalent circuit. (b) A typical Nyquist plot of electrochemical impedance spectroscopy (EIS).

3. $[Ru(bpy)_2DPPZ]^{2+}$ in the electrochemistry; search for the mediator of PCR

Our first step was to find a mediator molecule that would act as a mediator and intercalator of dsDNA simultaneously. As the PCR progresses, the dsDNA intercalates the mediator, resulting in an increase in Rct and allowing the PCR to be monitored in real time. We have selected $[Ru(bpy)_2DPPZ]^{2+}$, a derivative of $[Ru(bpy)_3]^{2+}$ that interacts with dsDNA and acts as a mediator. $[Ru(bpy)_2DPPZ]^{2+}$ was reported as a strong intercalator [33]. It binds mainly to the small groove of dsDNA, plugging the DPPZ portion between planar DNA base pairs [34, 35]. $[Ru(bpy)_2DPPZ]^{2+}$ also acts as an optical switch for double helical DNA with high luminescence sensitivity [33, 36]. It has therefore been studied in detail in chemosensors for the detection of luminescent signals. However, there have been few reports on the electrochemical properties.

First, cyclic voltammogram (CV) measurements were performed to investigate the electrochemical application of $[Ru(bpy)_2DPPZ]^{2+}$. **Figure 2** shows the measurement setup (Gamry. Instruments). The working electrodes of glassy carbon (GC) were polished with 1 μm of diamond paste and 50 nm aluminum oxide particles for 5 min each and then sonicated in ethanol and pure water for 5 min each. Pt wire was used for CE. The scan speed was 100 mV/sec and the potential range was -1 V to 1 V. (The same setting was used for the following measurements.) A solution of 10 mM Tris (pH 8.0) + 50 mM KCl + 1.5 mM $MgCl_2$ (PCRi) was used as the basic solution. **Figure 3(a)** shows the CV measurement result when 1 mM $[Ru(bpy)_2DPPZ]^{2+}$ was added. The redox peaks of $[Ru(bpy)_2DPPZ]^{2+}$ were observed in the negative potential region. The potential difference of the redox peaks (ΔEp) was large, suggesting that good charge transfer was not expected. For comparison, the CV, when 1 mM $[Fe(CN)_6]^{3-/4-}$ was added to the PCRi solution is shown in **Figure 3(b)**. The ΔEp was about 170 mV, suggesting that charge transfer is sufficiently rapid.

Next, EIS measurements were performed with 10 mVac and 1 kHz ~ 0.1 Hz in the PCRi solution. (The same conditions were used for the following EIS experiments). **Figure 3(e)** shows the results. The spectrum was a line extending almost vertically. The line was determined to be part of a semicircle with an extremely large Rct, and the Rct could be at least several hundred kΩ. This result indicated that the mediator function of $[Ru(bpy)_2DPPZ]^{2+}$ was low, which was initially expected.

Since 1 mM $[Ru(bpy)_2DPPZ]^{2+}$ alone did not provide sufficiently good charge transfer, we investigated whether the addition of 1 mM $[Fe(CN)_6]^{3-/4-}$, to this solution would lower the Rct resistance. **Figure 3(c)** shows the CV measurement results. The

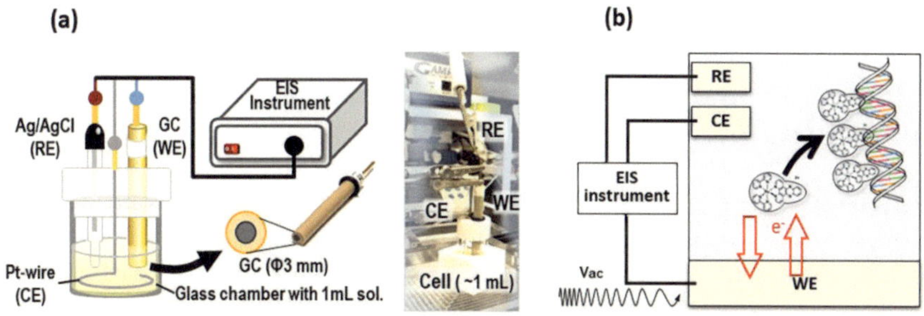

Figure 2.
(a) Schematic drawings of electrochemical measurement setting (3-electrode system) and a photo. (b) Desirable $[Ru(bpy)_2DPPZ]^{2+}$ function as a mediator and intercalator.

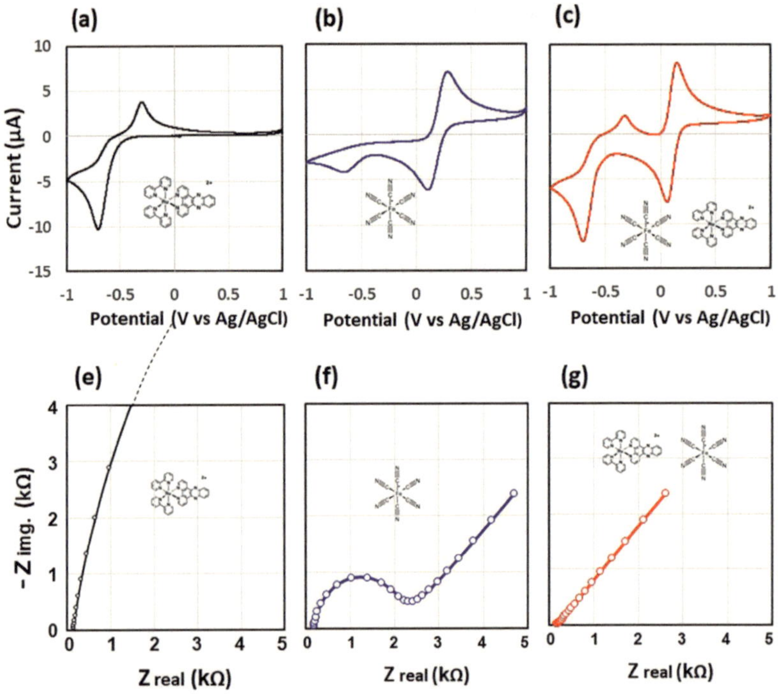

Figure 3.
CV and EIS results were obtained using PCRi solution with (a)(e) 1 mM $[Ru(bpy)_2DPPZ]^{2+}$, (b)(f) 1 mM $[Fe(CN)_6]^{3-/4-}$, (c)(g) both ions. The size of the CG-WE was ϕ3 mm.

obtained CV characteristics were similar to the sum of the peak position and current for 1 mM $[Ru(bpy)_2DPPZ]^{2+}$ and 1 mM $[Fe(CN)_6]^{3-/4-}$, except that the redox peaks of $[Fe(CN)_6]^{3-/4-}$ showed a substantial change. The redox peak current was larger and ΔEp was much narrower, about 90 mV. The result suggested that the charge transfer of $[Fe(CN)_6]^{3-/4-}$ mediator was enhanced by 1 mM $[Ru(bpy)_2DPPZ]^{2+}$ significantly. EIS measurements also confirmed this charge transfer enhancement (**Figure 3(f)**, (**g**)). The spectra showed that the semicircle in **Figure 3(f)** decreased to the extent that the semicircle is almost unobservable (**Figure 3(g)**). Rct decreased significantly.

From these measurements, we concluded that $[Ru(bpy)_2DPPZ]^{2+}$ alone cannot be used as a mediator for EIS measurements. However, there was a new finding that $[Ru(bpy)_2DPPZ]^{2+}$ greatly enhanced the charge transfer of $[Fe(CN)_6]^{3-/4--}$.

4. Qualitative evaluation of charge transfer enhancement by $[Ru(bpy)_2DPPZ]^{2+}$

We studied the dependence of charge transfer enhancement on the concentration of $[Ru(bpy)_2DPPZ]^{2+}$ by EIS. EIS measurements were repeated every 5 min and for the first 60 min, we waited for the equilibration, "termination" of the glassy carbon WE in a PCRi +1 mM $[Fe(CN)_6]^{3-/4-}$. After stabilization, $[Ru(bpy)_2DPPZ]^{2+}$ was added stepwise every 30 min to a final concentration of 1, 3, 5, 10, 30, 50, 100, 300, 500 nM, 1, 2, 3, 5, 10 μM, and 1 mM. After each addition, EIS measurements were performed every 5 min. All obtained EIS spectra were then fitted with a Randle equivalent circuit, and Rcts were

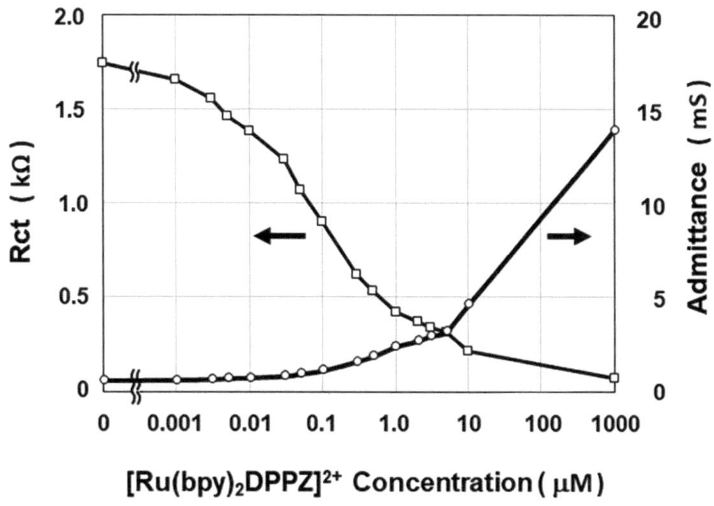

Figure 4.
The dependence of Rct and 1/Rct on [Ru(bpy)$_2$DPPZ]$^{2+}$ concentration. The charge transfer resistances were extracted from EIS spectra.

calculated using software (Gamry. Instruments). The obtained Rct and 1/Rct were plotted against [Ru(bpy)$_2$DPPZ]$^{2+}$ concentration in **Figure 4**.

Rct started to decrease from 1 nM concentration. At 100 nM concentration, Rct was reduced by half and about 1/4 at 1 µM. The Rct continued to decrease to approximately 1/25 at 1 mM concentration. In terms of admittance, Rct increases from 0.57 mS to 13.9 mS, suggesting that this phenomenon is not saturated even at concentrations above 10 µM. This result indicates that the charge enhancement effect of the [Ru(bpy)$_2$DPPZ]$^{2+}$ occurs over a wide range of concentrations from 1 nM to 1 mM, when the concentration of the [Fe(CN)$_6$]$^{3-/4-}$ mediator is 1 mM.

5. Charge transfer enhancement arises from the combination of [Ru(bpy)$_2$DPPZ]$^{2+}$, [Fe(CN)$_6$]$^{3-/4-}$ and carbon electrode

The Ru effect was found to occur with very small amounts of Ru complex ions, but the base solution contained 10 mM Tris pH 8.0, 50 mM KCl, and 1.5 mM $MgCl_2$.

We first removed the ionic substances 50 mM KCl and 1.5 mM $MgCl_2$ from the solution and performed EIS measurements. After stabilization, 0.5 µM and 1.0 µM [Ru(bpy)$_2$DPPZ]$^{2+}$ were added in sequence and EIS measurements were performed. **Figure 5(a)(e)** shows the change in EIS spectra and Rct decrease after the addition of [Ru(bpy)$_2$DPPZ]$^{2+}$. Upon addition of 0.5 µM [Ru(bpy)$_2$DPPZ]$^{2+}$, the Rct decreased rapidly and the Rct further decreased to almost 1/15 with f 1.0 µM. This result indicates that these KCl and $MgCl_2$ ions are not related to the charge transfer enhancement effect.

Next, the buffer, 10 mM Tris pH 8.0 was replaced with 10 mM phosphate buffer pH 8.0, and similar EIS measurements were performed. **Figure 5(b)(f)** shows the results. The addition of [Ru(bpy)$_2$DPPZ]$^{2+}$ rapidly decreased Rct to about 1/15 at 1 µM, similar to the Tris buffer, indicating that Tris buffer did not have the effect either.

To see the effect of pH, we also performed an experiment using 10 mM phosphate buffer pH 6.5 + 1.0 mM [Fe(CN)$_6$]$^{3-/4-}$. **Figure 5(c)(g)** shows the results. As in the case of pH 8.0, a sharp decrease in Rct was observed.

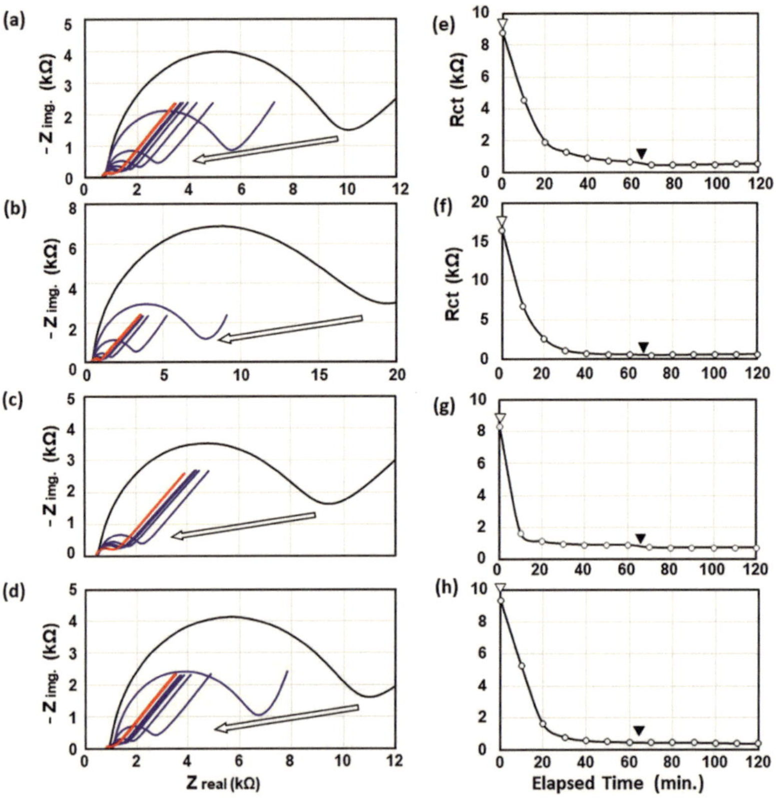

Figure 5.
EIS spectra and time-dependency of Rct without (black), with 0.5 μM (blue) and 1.0 μM (red) $[Ru(bpy)_2DPPZ]^{2+}$ in (a) (e)10 mM Tris pH 8.0 and 1.0 mM $[Fe(CN)_6]^{3-/4-}$, (b) (f) 10 mM phosphate buffer pH 8.0 and 1.0 mM $[Fe(CN)_6]^{3-/4-}$, and (c) (g) 10 mM phosphate buffer pH 6.5 and 1.0 mM $[Fe(CN)_6]^{3-/4-}$, (d) (h) 1.0 mM $[Fe(CN)_6]^{3-/4-}$. EIS spectra were drawn every 10 min. The white (△) and black (▲) mark the time points at which 0.5 μM and 1 μM of $[Ru(bpy)_2DPPZ]^{2+}$ were added, respectively.

Finally, the buffer was removed. EIS measurement was performed with only 1 mM$[Fe(CN)_6]^{3-/4-}$ in pure water. The pH of the solution was about 6.5. **Figure 5(d)(h)** shows the EIS results. Even in a simple solution of 1 mM $[Fe(CN)_6]^{3-/4-}$ alone, the addition of $[Ru(bpy)_2DPPZ]^{2+}$ ions promotes charge transfer, resulting in a significant decrease in Rct, which eventually reaches less than 1/20 of Rct.

Figure 5(e)-(h) shows the time dependence of Rct calculated from EIS measurements. In all cases, the Rct decrease was completed in about 20 minutes.

From these results, we can conclude that the newly discovered charge transfer enhancement effect is caused by the combination of $[Ru(bpy)_2DPPZ]^{2+}$, $[Fe(CN)_6]^{3-/4-}$, and a carbon electrode. The $[Ru(bpy)_2DPPZ]^{2+}$ enhances the charge transfer between the carbon electrode and $[Fe(CN)_6]^{3-/4-}$ by some mechanism. This effect is also observed at various carbon electrodes [37].

6. PCR monitoring by EIS

Based on the new findings, we conceived the idea of applying the charge transfer enhancement effect by the DNA intercalator to the real-time monitoring of PCR

[38]. In PCR, forward and revers primers of ssDNA are annealed to the DNA template, and DNA polymerase synthesizes dsDNA amplicons in heat cycles. Therefore, if $[Fe(CN)_6]^{3-/4-}$ and a small amount of $[Ru(bpy)_2DPPZ]^{2+}$ are added to the PCR solution and EIS measurement is performed using a carbon electrode, the progress of PCR can be monitored. The Rct would initially be low and as the PCR proceeds, the generated dsDNA would intercalate $[Ru(bpy)_2DPPZ]^{2+}$, and the Rct increases. The progress of PCR can be monitored by Rct.

To realize this monitoring method, we must first make a porous protective layer on the carbon electrode. We coated the carbon electrode with polyvinyl alcohol (PVA). Carbon electrodes were immersed in 0.1% PVA (30 k-51 kDa) for 1 hour and EIS measurements were performed in PCRi +1 mM $[Fe(CN)_6]^{3-/4-}$ solution, then 1 μM ssDNA (17 nt) and 1 μM dsDNA (60 bp) were added sequentially. The EIS spectrum remained almost unchanged from its initial shape, indicating that DNA adsorption was prevented (**Figure 6**).

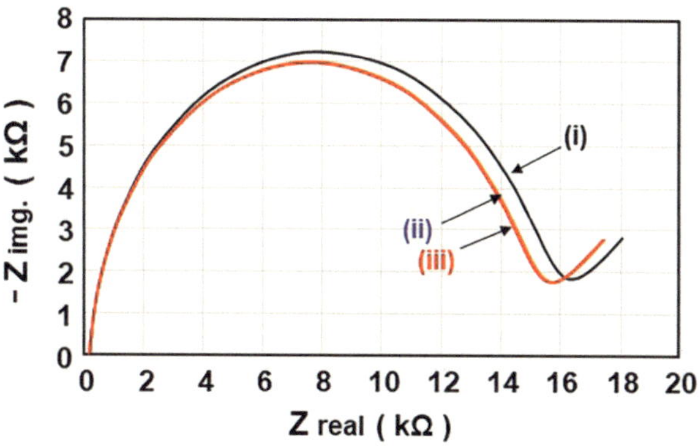

Figure 6.
EIS measurements using carbon electrode with PVA layer. (i) in PCRi + 1 mM, $[Fe(CN)_6]^{3-/4-}$ solution. (ii) after 1 μM ssDNA (17 nt) addition, (iii) after 1 μM dsDNA (60 bp) addition.

Integrating the results, PCR was performed; 3 mM $[Fe(CN)_6]^{3-/4-}$ and 5 μM $[Ru(bpy)_2DPPZ]^{2+}$ were added to the PCR solution, and PCR progress was monitored by EIS using a screen-printed carbon electrode. Template DNA concentrations were 100 ng/mL, 1 ng/mL, 10 pg/mL, and 0.1 pg/mL, and EIS measurements with a frequency range of 1 kHz to 0.5 Hz, open circuit potential, and 100 mVac were conducted during PCR extension steps.

All EIS spectra were fitted to a Randle equivalent circuit and Rcts were calculated. **Figure** 7(**a**) shows the change in Rct for each template DNA concentration versus heat cycle. Rcts were approximately 10–20 kΩ (See **Figure** 7(**d**)), however, its absolute value varied from sensor to sensor depending on the electrode preparation conditions, as is often the case [39]. To easily compare the results, all Rcts were normalized by setting the Rct at the eighth heat cycle as 1.0, and Rcts were overdrawn.

In the beginning, Rct increased gradually due to the adsorption of minute impurities on the electrode. Unlike this gradual increase, a clear abrupt increase in Rct occurred around 13, 21, 28, and 35 thermal cycles at DNA template concentrations of 100 ng/mL, 1 ng/mL, 10 pg/mL, and 0.1 pg/mL, followed by a gradual increase in Rct. This rapid increase in Rct is the result of the exponential amplification of the dsDNA amplicon and intercalation of $[Ru(bpy)_2DPPZ]^{2+}$.

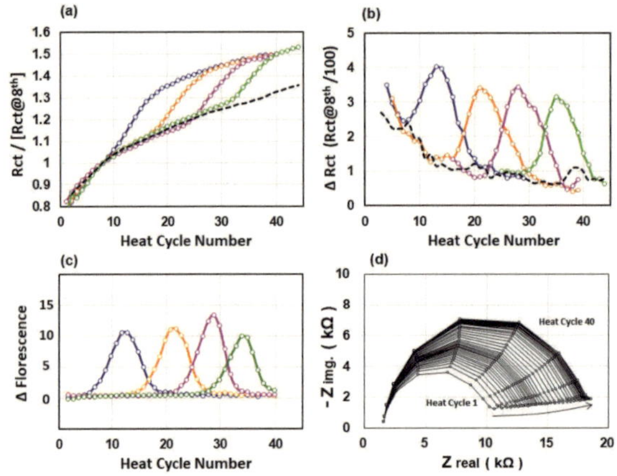

Figure 7.
Monitoring results of PCR with DNA template concentrations of 100 ng/mL, 1 ng/mL, 10 pg/mL, and 0.1 pg/mL (a) Rct calculated from EIS by fitting to the Randle circuit with constant phase element, (b) ΔRct; Rct increase at each heat cycle, and (c) optical fluorescence increase. Background without template DNA is shown as a broken black line. All data are three moving averages. (d) One example of changes in EIS spectra during PCR at the extension step. The Arrow indicates spectral changes along the heat cycle.

The change in Rct is more clearly confirmed by ΔRct, the increase in normalized Rct at each thermal cycle (**Figure 7(b)**). The vertical axis is 1/100th scale of the Rct@8th.

The more DNA template in the initial solution, the earlier by 7 cycles the heat cycle where the peak starts, increasing by seven cycles each. Theoretically, the number of amplicons would double with each cycle. The present results correspond well to the theoretical values. Gel electrophoresis of the PCR amplicon samples also confirmed that the target DNA was successfully amplified. It can be concluded that the inhibitory effect of the addition of 3 mM $[Fe(CN)_6]^{3-/4-}$ and 5 μM $[Ru(bpy)_2DPPZ]^{2+}$ is negligible. PCR monitoring with EIS was successfully demonstrated.

Optical PCR was performed in parallel using the same PCR solution (micPCR, Bio Molecular Systems, Australia). 20 μL of the solution was added with 1 U/L SYBR Gold (Toyobo Corporation), PCR was performed, and fluorescence was measured. **Figure 7(c)** shows the fluorescence intensity increase/cycle (Δfluorescence). Δflorescence showed a peak with 10 thermal cycle width. The peak positions by impedance and optical methods were in good agreement. The critical threshold cycle (Ct) was confirmed to be detectable by EIS.

The above results show that the electrochemical detection method which utilizes the anomalous charge transfer enhancement effect can monitor the progress of PCR in real time and realize quantitative PCR [38].

7. Conclusion

We proposed a new impedance detection method for real-time PCR based on EIS and demonstrated its practicality. The simplicity of this impedance detection method allowed us to fabricate qPCR with simple components and to realize a small, portable,

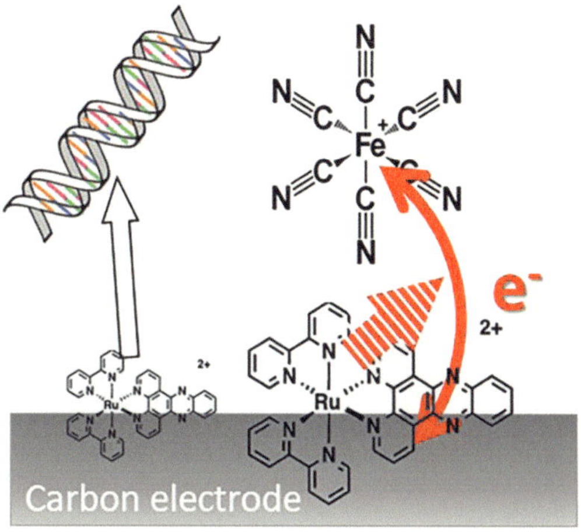

Figure 8.
A possible charge transfer enhancement mechanism and the relation with $[Ru(bpy)_2DPPZ]^{2+}$ *and dsDNA.* $[Ru(bpy)_2DPPZ]^{2+}$ *adsorbs on the carbon electrode and enhances the charge transfer through the ligands.*

robust, and inexpensive device. At present, the mechanism of this effect is not clear. It was known that $[Fe(CN)_6]^{3-/4-}$ is a surface-sensitive mediator, therefore, the effect may have been caused by interaction or specific adsorption of $[Ru(bpy)_2DPPZ]^{2+}$ on the basal or edge surfaces of the carbonaceous electrode, which facilitates charge transfer (see **Figure 8**). This effect is observed in a variety of carbon electrodes, including graphite, graphene, and screen-printed carbon, and is expected to be widely applied beyond this PCR application.

Acknowledgements

This research is partially supported by the Center of Innovation Program (Grant No. JPMJCE1310) and Core Research for Evolutional Science and Technology, CREST (Grant No. JPMJCR18I3) from Japan Science and Technology Agency, JST.

Conflict of interest

The authors declare no conflict of interest.

Author details

Huanwen Han[1], Kazuyuki Nobusawa[1,2], Fumie Takei[3], Ting-Chieh Chu[1], Noriyasu Hashida[4] and Ichiro Yamashita[1*]

1 Graduate School of Engineering, Osaka University, Suita, Osaka, Japan

2 Textile Research Institute of Gunma, Kiryu, Japan

3 National Defense Medical College, Tokorozawa-shi, Saitama, Japan

4 Department of Ophthalmology, Osaka University Graduate School of Medicine, Suita, Osaka, Japan

*Address all correspondence to: yamashita@jrl.eng.osaka-u.ac.jp

References

[1] Huang Y, Xu J, Liu J, Wang X, Chen B. Disease-related detection with electrochemical biosensors: A review. Sensors. 2017;**17**(10):2375

[2] Naveen MH, Gurudatt NG, Shim YB. Applications of conducting polymer composites to electrochemical sensors: A review. Applied Materials Today. 2017;**9**:419-433

[3] Shao Y, Wang J, Wu H, Liu J, Aksay IA, Lin Y. Graphene based electrochemical sensors and biosensors: A review. Electroanalysis. 2010;**22**(10):1027-1036

[4] Wang Z, Dai Z. Carbon nanomaterial-based electrochemical biosensors: An overview. Nanoscale. 2015;**7**:6420-6431

[5] Barsan MM, Ghica ME, Brett CMA. Electrochemical sensors and biosensors based on redox polymer/carbon nanotube modified electrodes: A review. Analytica Chimica Acta. 2015;**881**:1-23

[6] Zhu C, Yang G, Li H, Du D, Lin Y. Electrochemical sensors and biosensors based on nanomaterials and nanostructures. Analytical Chemistry. 2015;**87**(1):230-249

[7] Saidur MR, AbdulAziz AR, Basirunbc WJ. Recent advances in DNA-based electrochemical biosensors for heavy metal ion detection: A review. Biosensors & Bioelectronics. 2017;**90**:125-139

[8] Li H, Liu X, Mu X, Genov R, Mason AJ. CMOS electrochemical instrumentation for biosensor microsystems: A review. Sensors. 2017;**17**(1):74

[9] Fujino Y, Nakamura R, Han H-W, Yamashita I, Shimizu T, Shingubara S, et al. Electrochemical impedance spectroscopy study of liposome adsorption and rupture on self-assembled monolayer: Effect of surface charge. Electroanalytical Chemistry. 2020;**878**:114572

[10] Faye O, Faye O, Diallo D, Diallo M, Weidmann M, Sall AA. Quantitative real-time PCR detection of Zika virus and evaluation with field-caught Mosquitoes. Virology Journal. 2013;**10**:311

[11] Park M, Won J, Choi BY, Lee CJ. Optimization of primer sets and detection protocols for SARS-CoV-2 of coronavirus disease 2019 (COVID-19) using PCR and real-time PCR. Experimental & Molecular Medicine. 2020;**52**:963-977

[12] Lehmann LE, Hunfeld K-P, Emrich T, Haberhausen G, Wissing H, Hoeft A, et al. A multiplex real-time PCR assay for rapid detection and differentiation of 25 bacterial and fungal pathogens from whole blood samples. Medical Microbiology and Immunology. 2008;**197**;313-324

[13] Kralik P, Ricchi M. A Basic Guide to Real Time PCR in microbial diagnostics: Definitions, parameters, and everything. Frontiers in Microbiology. 2017;**8**. Article 108

[14] Resnick KE, Alder H, Hagan JP, Richardson DL, Croce CM, Cohn DE. The detection of differentially expressed microRNAs from the serum of ovarian cancer patients using a novel real-time PCR platform. Gynecologic Oncology. 2009;**112**:55-59

[15] Spurgeon SL, Jones RC, Ramakrishnan R. High throughput gene expression measurement with real time

PCR in a microfluidic dynamic array. PLoS One. 2008;**3**:e1662

[16] Myakishev MV, Khripin Y, Hu S, Hamer DH. High-throughput SNP genotyping by Allele-specific PCR with universal energy-transfer-labeled primers. Genome Research. 2001;**11**:163-169

[17] Yeung SSW, Lee TMH, Hsing I-M. Electrochemistry-based real-time PCR on a microchip. Analytical Chemistry. 2008;**80**:363-368

[18] Defèver T, Druet M, Dequaire MR, Joannes M, Grossiord C, Limoges B, et al. Real-time electrochemical monitoring of the polymerase chain reaction by mediated redox catalysis. American Chemical Society. 2009;**131**:11433-11441

[19] Defèver T, Druet M, Evrard D, Marchal D, Limoges B. Real-time electrochemical PCR with a DNA intercalating redox probe. Analytical Chemistry. 2011;**83**:1815-1821

[20] Moreau M, Delile S, Sharma A, Fave C, Perrier A, Limoges B, et al. Detection of a few DNA copies by real-time electrochemical polymerase chain reaction. Analyst. 2017;**142**:3432-3440

[21] Chen J, Yang T, Jiao K, Gao H. A DNA electrochemical sensor with poly-L-lysine/single-walled carbon nanotubes films and its application for the highly sensitive EIS detection of PAT gene fragment and PCR amplification of NOS gene. Electrochimica Acta. 2008;**53**:2917-2924

[22] Wang L, Liu Q, Hu Z, Zhang Y, Wu C, Yang M, et al. A novel electrochemical biosensor based on dynamic polymerase-extending hybridization for E. coli O157:H7 DNA detection. Talanta. 2009;**78**:647-652

[23] Corrigan DK, Schulze H, Henihan G, Ciani I, Giraud G, Terry JG, et al. Impedimetric detection of single-stranded PCR products derived from methicillin resistant Staphylococcus aureus (MRSA) isolates. Biosensors & Bioelectronics. 2012;**34**:178-184

[24] Chen C-C, Lai Z-L, Wang G-J, Wu C-Y. Polymerase chain reaction-free detection of hepatitis B virus DNA using a nanostructured impedance biosensor. Biosensors & Bioelectronics. 2016;**77**:603-608

[25] Lasia A. In: Conway BE, Bockris JOM, White RE, editors. Modern Aspects of Electrochemistry No.32. Electrochemical Impedance Spectroscopy and its Applications. Boston, MA: Springer; 2002. pp. 143-248. DOI: 10.1007/0-306-46916-2_2

[26] Bogomolova A, Komarova E, Reber K, Gerasimov T, Yavuz O, Bhatt S, et al. Challenges of electrochemical impedance spectroscopy in protein biosensing. Analytical Chemistry. 2009;**81**(10):3944-3949

[27] Chang BY, Park SM. Electrochemical impedance spectroscopy. Annual Review of Analytical Chemistry. 2010;**3**:207-229

[28] Lisdat F, Schäfer D. The use of electrochemical impedance spectroscopy for biosensing. Analytical and Bioanalytical Chemistry. 2008;**391**:155

[29] Park JY, Park SM. DNA hybridization sensors based on electrochemical impedance spectroscopy as a detection tool. Sensors. 2009;**9**(12):9513-9532

[30] Randviir EP, Banks CE. Electrochemical impedance spectroscopy: An overview of bioanalytical applications. Analytical Methods. 2013;**5**:1098-1115

[31] Grossi M, Riccò BJ. Electrical impedance spectroscopy (EIS) for biological analysis and food characterization: A review. Journal of Sensors and Sensor Systems. 2017;**6**:303-325

[32] Li X, Zhang Y, Chen H, Sun J, Feng F. Protein nanocages for delivery and release of luminescent ruthenium (II) polypyridyl complexes. ACS Applied Materials & Interfaces. 2016;**8**:22756-22761

[33] Friedman A, E, Chambron J-C, Sauvage J-P, Turro NJ, Barton JK. Molecular "Light Switch" for DNA: $Ru(bpy)_2(dppz)^{2+}$. Journal of the American Chemical Society. 1990;**112**:4960-4962

[34] Li G, Sun L, Ji L, Chao H. Ruthenium (II) complexes with dppz: From molecular photoswitch to biological applications. Dalton Transactions. 2016;**45**:13261-13267

[35] Song H, Kaiser J, Barton JK. Crystal structure of Δ-$[Ru(bpy)_2dppz]^{2+}$ bound to mismatched DNA reveals side-by-side metalloinsertion and intercalation. Nature Chemistry. 2012;**4**:615-620

[36] Brennaman MK, Alstrum-Acevedo JH, Fleming CN, Jang P, Meyer TJ, Papanikolas JM. Turning the $[Ru(bpy)_2dppz]^{2+}$ light-switch on and off with temperature. Journal of the American Chemical Society. 2002;**124**:15094-15098

[37] Han H, Nobusawa K, Yamashita I. Anomalous enhancement of electrochemical charge transfer by a Ru complex ion intercalator. Analytical Chemistry. 2022;**94**(2):571-576

[38] Nobusawa K, Han H, Takei F, Chu T, Hashida N, Yamashita I. Electrochemical impedimetric real-time polymerase chain reactions using anomalous charge transfer enhancement. Analytical Chemistry. 2022;**94**(22):7747-7751

[39] McCreery RL. Advanced carbon electrode materials for molecular electrochemistry. Chemical Reviews. 2008;**108**:2646-2687

Chapter 2

Leveraging Dendrimer Macromolecules for the Encapsulation and Stabilisation of Nano-Sized Ruthenium Catalysts: Evaluation of Catalytic Reaction Kinetics in the Reduction of Pollutants Organic Dyes, Oxidation of Alcohols and Alkenes as Well as Hydrogenation Reactions

Mulisa Maumela and Ndzondelelo Bingwa

Abstract

Encapsulation of nano-sized metal catalysts within the dendrimers macromolecules' frameworks has been well documented thus far. Dendrimers are described as symmetric, monodispersed macromolecules resembling a tree-like branched structure and have been utilised as both a template and stabilising agent for the fabrication of metal (noble and non-noble) nano-catalysts. For this purpose, different types of dendrimers can be employed. The use of dendrimers for metal catalysts stabilisation or encapsulation offers several advantages in catalysis. For example, the dendrimer template allows the synthesis of catalytically active monodispersed nanoparticles and the dendrimers template itself does not passivate the metal active atoms during the catalytic process. Additionally, dendrimers have the potential to act as a "vehicle" that can be leveraged for the fabrication of heterogeneous catalysts. For example, surface groups of the dendrimers can be functionalised to chemically link the dendrimer-encapsulated nanoparticles (DENs) with solid supports such as silica. A significant number of studies on the synthesis and catalytic evaluation of dendrimer-metal nanocomposite materials (e.g. Ruthenium-based) onvarious reactions can be found in the literature. This chapter, however, will particularly focus on the recent developments on the synthesis, characterisation and catalytic applications of dendrimer-derived (colloidal and supported) Ruthenium catalysts.

Keywords: dendrimers, ruthenium, dendrimer-encapsulated nanoparticles (DENs), catalysis, hydrogenation, reduction, oxidation

1. Introduction

With the rapid increase in environmental problems such as pollution and climate change, the search for chemical processes that offer solutions to mitigate these challenges has also intensively received a lot of attention from researchers around the globe in the past decades. More recently, the revised Transformation Agenda by the United Nation has firmly advocated for the attainment of Sustainable Development Goals (SDG) by year 2030 [1]. As a result, more efforts have been projected into the development of sustainable material science and nanotechnology. For example, it has become more imperative to innovate materials and technology that can be utilized in the nanoscale range. Additionally, the materials and technologies must be environmentally friendly, recyclable, cost-effective and sustainable. The synthesis of nano-sized metal and non-metal catalysts using "green" processes, among other chemical processes, is one of the examples of the ongoing efforts by scientists towards the attainment of SDG. This is because nanocatalysts derived from metal nanoparticles have been found to possess different and unique chemical and physical characteristics as compared to those observed in their bulk counterpart catalysts [2]. Some of these characteristics include higher surface area, robustness, and better stability as compared to other conventional nanocatalysts [3–5]. Due to the high surface area of metal nanoparticles, a significant portion of their active atoms are available for catalysis. This in turn, results in the superior performance of nanoparticle catalysts due to enhanced contact between the reactants and the catalysts during the reaction process. It is also widely believed that as the size of the metal NPs decreases, the surface area increase (and vice-versa) and that the catalytic rate is directly proportional to the number of catalytic active sites exposed [6].

Because metal nanoparticle catalysts are well dispersed in solution, particularly unsupported colloidal nanoparticles, it has always been tempting to often classify them as "homogeneous" catalysts. However, these metal NPs have been perceived to be insoluble in many organic solvents use for common yet important reactions. On assumption that colloidal NPs do not dissolve in organic solvents that are often used during chemical reactions, they are rendered to act as "heterogeneous green' catalysts [7]. As such, it is believed that these colloidal NPs can be recovered easily after the reaction using suitable various techniques. However, the recovery and separation of colloidal nanoparticles from the final reaction products have been found to be a complex process and hence very little progress has been made on the advancement of recycling these catalysts to this end [8–10].

Some attempts to establish whether nanoparticle catalysed reactions follow a homogeneous or heterogeneous mechanisms have been made by few researchers. For example, in a study published by Nemanashi and Meijboom entitled; "cat in a bag recycling of dendrimer encapsulated Au nanoparticles by use of dialysis membrane bag in the reduction of 4-nitrophenol reduction: proof of heterogeneous catalysis", a heterogeneous type of mechanism was suggested [11]. In their study, dendrimer encapsulated Au nanoparticles Au-DENs with an average particle size ranging between 2.1 and 3.1 nm were trapped into a permeable dialysis membrane bag which only allows for the diffusion (in and out) of small reactants molecules. The entrapped Au-DENs were dipped into a reaction flask containing aqueous solutions of 4-Nitrophenol and an excess reducing agent ($NaBH_4$) while keeping the system under stirring. The reactants molecules diffused inside the dialysis bag containing the catalyst and made contact with the surface of the Au-DENs catalyst for the product formation. The product (4-aminophenol) formation was monitored by analysing a reaction solution sampled over time using UV-vis spectrometry. Post-reaction characterisation of the catalyst by techniques such as Transmission Electron Microscope (TEM) revealed that the DEN catalysts retained their original particle sizes and showed no leaching even after three reaction cycles. It was concluded that the catalytic reduction of 4-NP occurs on the NPs' surface as opposed to induced by leaching of metal NPs atoms into solution. However, some researchers have suggested in their review articles that both heterogeneous and homogeneous type of mechanisms for various metal nanoparticles catalysed reactions [12, 13]. Narayan et al. have recently reviewed the synthesis and utilisation of metal nanoparticles (NPs) as "green" catalysts [7]. The issue of whether metal nanoparticles catalysis follows a homogeneous or heterogeneous reaction mechanism remain an ongoing debate in catalysis research.

2. Synthesis of metal nanoparticles

Generally, two main approaches are employed for the synthesis of nanoparticles—top-to-bottom method and the bottom-up [14]. The main distinct difference between these two methods is the starting precursor for the formation of nanoparticles. For instance, the top-to- bottom approach entails the process in which the bulk material is used as a starting material and the particle size is reduced to nano-sized particles via different physical, chemical chapter, or mechanical processes [15]. On the other hand, the bottom-up approach involves making use of molecules, small particles or atoms as starting materials for the assembling of nanoparticles via processes such as the liquid state synthesis method (chemical reduction) [16]. This particular chapter will solely focus on the latter approach for the synthesis of nanoparticle with more emphasis on the synthesis of dendrimer stabilised Ruthenium nanoparticle catalysts. The synthesis of dendrimer-based Ruthenium nanocatalysts (Ru-DENs) will be discussed in detail in the succeeding sections of this chapter.

Based on the afore-mentioned main two approaches that are used for the preparation of metal nanoparticles, different methods (physical and chemical processes) can be explored for their synthesis depending on the targeted final physical and chemical properties (size, morphology, application, etc) of the nanoparticles to be produced. For instance, a physical process such as flame spray pyrolysis (FSP) can be employed if the target is to produce nanoparticles on a large scale [17]. Alternatively, the synthesis of monodispersed nanoparticles with a well-defined spherical shape can be achieved by employing a chemical process proposed by J. Turkevich. For example, Kimling et al. reported the synthesis of spherical, monodispersed Au nanoparticles using the Turkevich method [18]. This was simply achieved by reduction of citrate at 100°C, and subsequent mixing with gold hydrochlorate solution. To synthesise nanoparticles of different sizes using this method, the concentration of citrate solution (capping agent) can manipulated.

Equally so, a variety of stabilising or capping agents can be employed depending on the method chosen for the synthesis of metal nanoparticles. The use of a suitable stabiliser during the synthesis of metal nanoparticles is vital, particularly those aimed to be utilised in catalysis. This stabilisation is primarily vital to circumvent unwanted phenomena to occur on the catalyst, particularly during the catalytic processes usually carried out under harsh conditions. For example, phenomena such as sintering, migration-coalescence, and agglomeration may lead to catalyst deactivation during the catalysis process. Therefore, various organic ligands such as ionic liquids (ILs) [2, 19–22], simple amines, phosphines, and thiols [12, 23, 24], cyclodextrins [25–29], caliraxenes [30–32], polymers [7, 12, 33–35], surfactants [36, 37], and dendrimers [38–43]. Of these metal nanoparticles' stabilising agents, dendrimers have been found to be versatile and potentially offer a considerable number of advantages, particularly in catalysis [40, 43]. This chapter gives a comprehensive discussion on the utilisation of dendrimers macromolecules as stabilising and encapsulating agents for ruthenium metal nanoparticles as well as their applications in various reactions such as reduction of pollutants dyes, oxidation of methanol, hydrogenations of biomass-derived molecules and other organic compounds.

Dendrimers macromolecules are classified as hyperbranched polymers, with a well-defined, monodispersed, tree-like three-dimensional structure. These versatile macromolecules possess three main structural components; (a) interior core, (b) interior branching units, and (c) the periphery, attached to the outermost branching unit (see **Figure 1**). The synthesis of dendrimers is carried out using either the convergent or the divergent methods. The divergent approach entails an iteration sequence of reaction steps in which iteration from the core unit lead to higher dendrimer generation. On the contrary, the divergent method involves a process of synthesising the dendrimer from periphery units towards the core units via a two-stage process. Vögtle et al. reported the first synthetic procedure for the construction of well- defined dendrimers in 1978 via divergent approach and referred to the used procedure as "cascade synthesis" [45]. Few years later, Tomalia et al. reported the modified Vögtle's procedure for the synthesis of independent, divergent, and macromolecular "true dendrimers", poly(amidoamine) (PAMAM) dendrimers [46]. Newkome et al. also reported the synthesis of what they termed "arborols" (a synonym for dendrimers) during the same year as Tomalia and co-workers [47]. The first example for the convergent synthesis of dendrimers was reported by Hawker and Fréchet in 1990 [48].

Since their (dendrimers) discovery, these materials have found a wide range of applications in nanoscience. For instance, they have extensively been used in areas such as in drug delivery [49], sensors [50], electronic devices [51], wastewater treatment [52] and catalysis [53], to name just a few. However, this chapter will solely focus on the latter applications, in which dendrimers are utilised for the stabilisation and

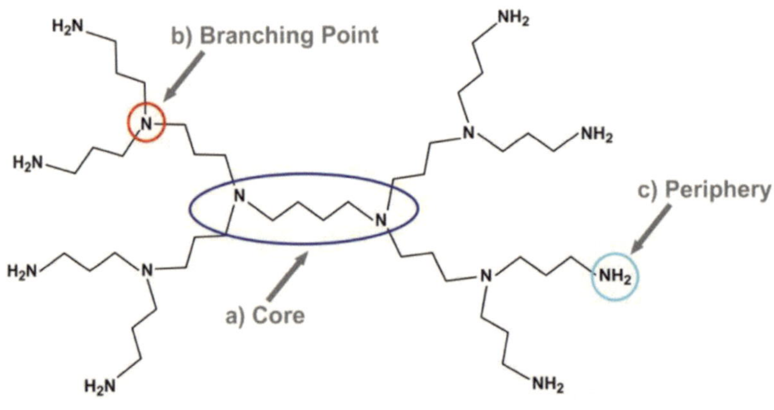

Figure 1.
A typical schematic illustration of the structural components of poly(propylene imine) (PPI) dendrimer molecule [44].

encapsulation during the synthesis of catalytically active homogeneous and heterogeneous metal nanoparticles, particularly ruthenium based.

3. Dendrimers for encapsulation of metal nanocatalysts

Green chemistry advocates for the use of set of principles that seek to promote elimination or minimising the use or generation of hazardous chemicals during the design, manufacture, and application of chemical products. Catalysis is one of the outlined important principles (of the 12) of green chemistry. The evolution of nanomaterials (such as metal nanoparticles) in the past decades offers a great platform for the attainment of sustainable, green catalysis practises. Many essential products (such as medicine, polymers, fibres, fuels, paints, lubricants) used by humans in the modern time are produced through catalytic processes. Therefore, catalysis plays a pivotal role in chemical transformation in chemical industries. For example, it has been reported that 90% of manufactured chemicals involves at least one catalytic process [54]. Because nanocatalysts possess high surface energy, which renders them thermodynamically unstable, possibilities of deactivation due to migration and coalescence during catalytic processes are always high, especially for processes carried out at elevated temperatures.

As a strategy for the fabrication of more stable, sustainable and catalytically active "green" metal nanocatalysts, many researchers have explored the encapsulation of metal nanocatalysts within different materials (encapsulating agents) using various methods. These encapsulating agents include materials such as inorganic oxides nanoshells (SiO_2, TiO_2, CeO_2), nanoporous materials (mesoporous materials, zeolites, metal-organic frameworks, covalent organic frameworks), and organic capsules (dendrimers, porous organic cages) [55]. In most cases, the synergetic effect is established between the metal nanoparticles and the encapsulating material and has been observed to greatly enhance the catalyst's activity and selectivity. More importantly, such catalysts are more stable and can easily be recyclable without losing their activity, making them almost ideal "green" catalysts. This chapter will focus on the encapsulation of ruthenium nanocatalysts within dendrimer frameworks (Ru-DENs), their characterisation, and catalytic applications.

4. Synthesis and characterisation of dendrimer-encapsulated ruthenium as well as their applications in catalysis

A simple two-step process involving complexation of the aqueous dendrimer solution and metal ions, followed by the reduction of the dendrimer-metal ions composite precursors is generally carried out for the synthesis of DENs. The dendrimer-metal ion composite, a precursor for DENs, is prepared by addition of the desired amount of metal ions (such as Pt^{2+}, Pd^{2+}, Au^{3+}, Ag^{+}, Cu^{2+}, Ni^{2+}, Ru^{3+}) to the dendrimer solution [43]. This process is usually carried out in aqueous media, however, in some cases, organic solvents have been used [56]. Different generations of PAMAM dendrimers having complexing (−NH) and non-complexing (−OH) peripheral groups have been used for this purpose. It is noteworthy to mention that other types of dendrimers have also been used successfully for the synthesis of DENs [57]. Metal ions partition into the interior tertiary amine groups of the dendrimer for a complete complexation of the dendrimer-metal ions composite. The extent of this complexation process can be monitored using simple characterisation techniques such as UV-vis, EPR, MALDI-MS, and XPS [58]. For example, during the complexation of PAMAM dendrimers with Cu^{2+} ions for the preparation of Cu-DENs, a strong ligand-to-ligand charge transfer (LMCT) band at λ 300 nm and a 605 nm shift of the d-d optical transition is observed [59]. Subsequent addition of an excess reducing agent (such as $NaBH_4$ or H_2) to the dendrimer-metal ion composite result in the formation of mono-dispersed small nanoparticles trapped within the dendrimer cavities/framework. **Figure 2** gives an example of a schematic illustration for the synthesis of dendrimer encapsulated mono- and bimetallic Ru nanoparticles. Various synthetic methods for the preparation of bimetallic DENs are detailed elsewhere [42, 60].

In cases where dendrimers used for the complexation of metal ions possess primary groups, such as amine-terminated dendrimers, the reduction process may lead to the formation of nanoparticles that are not entirely encapsulated within the dendrimer frameworks. The coordination of metal ions to the primary amine groups in such instance can be avoided by carrying out the synthesis under acidic conditions. This is because, the commonly used dendrimers, PAMAM, are polyprotic bases possessing surface primary amine groups that are more basic than interior tertiary amine groups. As such, acidic conditions have been shown to effectively drive the coordination of metal ions to the tertiary amines while avoiding coordination with surface primary amine groups on the periphery [61].

There are several advantages outlined for the use of dendrimers as hosts materials for metal nanoparticles intended for utilisation in catalysis in various reactions: (i) because dendrimers themselves are of a uniform structure and composition, the yielded encapsulated nanoparticles are of well-defined shape [62], (ii) since the nanoparticles are encapsulated within the dendrimer frameworks, agglomeration, which may lead to deactivation during catalysis is greatly circumvented [63], (iii) the nanoparticles trapped within the dendrimer framework are confined primarily by steric effects and a significant portion of their active atoms is unpassivated and available to take part in catalytic reactions [10], (iv) dendrimer branches can be used as selective gates to allow access of substrates molecules to the catalytically active

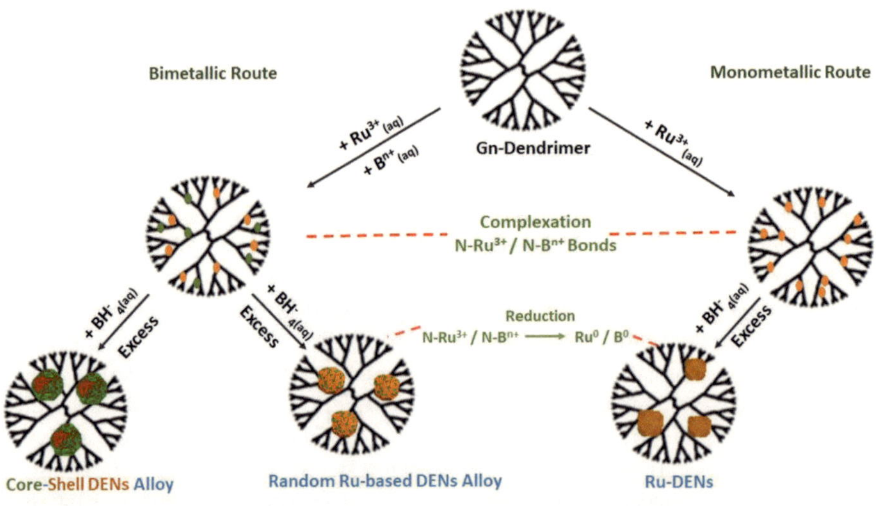

Figure 2.
Schematic illustration for the synthesis of monometallic and bimetallic Ru-DENs.

encapsulated metal nanoparticles [64], and (v) the primary groups on the periphery of the dendrimers can be tailored to control the solubility of the hybrid nanocomposite and used as a "vehicle" for linking nanoparticles to the surfaces and other polymers for "heterogenization" of colloidal nanoparticles [65, 66].

With the advancement of technologies in recent years, these as-synthesised DENs catalysts have been characterised using more sophisticated techniques [67]. This characterisation is carried out to determine physical and chemical properties such as the actual size, surface structure, valency, chemical composition, electron band gap, light emission, absorption, and scattering and diffraction properties. The most commonly used characterisation techniques for this purpose include nuclear magnetic resonance spectroscopy (NMR), infra-red spectroscopy (IR), ultra-violet and visible spectroscopy (UV-vis), transition electron microscopy (TEM), scanning tunnelling microscopy (STM), scanning electron microscopy (SEM), energy dispersive X-ray spectrometer (RDS), X-ray diffraction (XRD), X-ray photoelectron spectroscopy (XPS), ultraviolet photoelectron spectroscopy (UPS), extended X-ray adsorption fine structure spectroscopy (EXAFS), X-ray absorption near-edge spectroscopy (XANES), atomic force microscopy (AFM).

Crooks and co-workers pioneered the synthesis of DENs. The first example of the synthesis of DENs was reported for Cu-DENs by the same research group of Crook [59]. Fourth-generation hydroxyl-terminated PAMAM dendrimers (G4-PAMAM-OH) was used as a templating agent for this purpose. UV-vis spectroscopy was used to monitor the coordination of Cu^{2+} ions with the tertiary amine groups of dendrimers. For example, the weak, broad absorption band responsible for aqueous Cu^{2+} ions at λ 810 nm significantly shifted to 605 nm and the emergence of a new strong band at 300 nm in the case of G4-PAMAM/Cu^{2+} composite solution. Titration experiments conducted for the coordination of Cu^{2+} ions to the tertiary amine groups of the dendrimers led to the conclusion that each Cu^{2+} ion is more likely to coordinate with two tertiary amine groups of the G4-PAMAM-OH dendrimers. Subsequent reduction of the G4-PAMAM/Cu^{2+} composite solution by excess $NaBH_4$ resulted in the formation of Cu-DENs having an average particle diameter of less than 1.8 nm as determined by TEM. The formation of these Cu-DENs was also evidenced by the disappearance of the initial bands of the G4-PAMAM/Cu^{2+} composite solution. Additionally, a change in solution colour from blue to golden brown colour was observed as an indication for the formation of Cu-DENs.

Since then, there has been a plethora of publications covering the use of dendrimers for the encapsulation of other noble and non-noble metals such as Pt, Au, Ag, Pd, Rh, and Ru [41, 67]. However, this chapter will exclusively put more focus on the synthesis of Ru-DENs (colloidal) and Ru-DENs derived supported Ru catalysts as well as their applications in various reactions such as hydrogenation of alkenes or carbonyl compounds, and oxidation of alcohols. Additionally, the bimetallic alloys of Ru-based DEN catalysts prepared will also be fully explored. Ruthenium based catalysts (supported or unsupported) were rarely used in the past to catalyse hydrogenation reactions involved in the petrochemical industry and chemical synthesis. Instead, optimum activities and selectivities of the desired products for these reactions were achieved by utilisation of other catalysts such as platinum, palladium, rhodium, or nickel. However, ruthenium catalysts have been found to exhibit superior activity in such reactions in the past few decades [68–70]. For example, a review article entitled "heterogeneous Ru catalysts as the emerging potential superior catalysts in the selective hydrogenation of bio-derived levulinic acid to γ-valerolactone: effect of particle size, solvent, and support on activity, stability, and selectivity" has been recently published by Maumela et al. [71].

The first example for the synthesis of dendrimer-encapsulated Ru nanoparticles was reported by Lafaye et al. [72]. Aqueous solutions of hydroxyl-terminated fourth generation PAMAM dendrimers and $RuCl_3$ were used for the preparation of the dendrimer/metal ion composite using a metal ion to dendrimer molar ratio of 40:1. The coordination of the Ru^{3+} ions to the dendrimer amine groups was monitored by UV-vis spectroscopy in this case. Absorption peaks at 286 and 400 nm were both attributed to the complexation of the Ru^{3+} ion and the tertiary amine groups within the PAMAM dendrimer template. However, when this complexation is allowed to occur for a lengthy period of time (3 days) under inert conditions, the shift in the initial peaks was observed. For example, the absorption peak initially at 286 nm shifted to 280 nm, while that at 400 nm shifted to 407 nm. These peaks were still assigned to the Ru-amine complexation that formed within the dendrimer structure. The reduction of the composite solution was carried out by bubbling H_2 for 1 h at room temperature, yielding nanoparticles with a narrow size distribution and an average diameter of 1.2 nm (see **Figure 3a**). These as-synthesised Ru-DENs were subsequently immobilised onto Al_2O_3 support *via* wet impregnation method, followed by the removal of the dendrimer template by thermal treatment at 300°C. The thermal treatment, however, resulted in an increase in particle size by almost 50%, with average size being measured to be 2.0 nm (see **Figure 3b**),

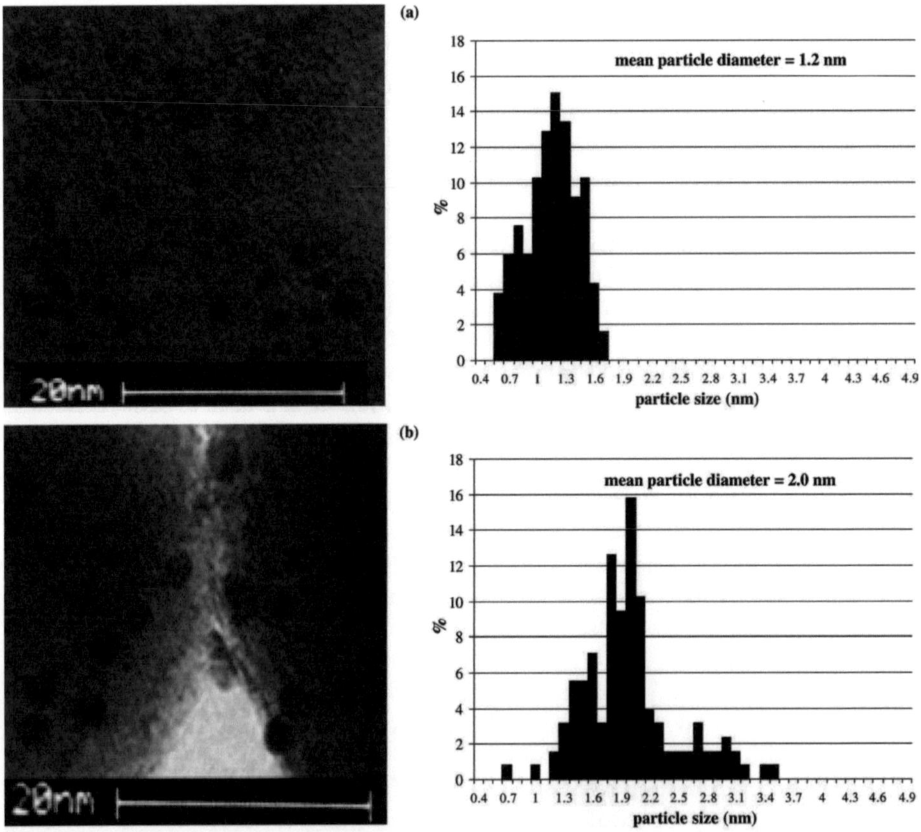

Figure 3.
HRTEM images and metal particle size distribution of (a) the Ru-G_4OH PAMAM dendrimer nanocomposite precursor, (b) a 1%wt Ru/Al_2O_3 catalyst prepared from the Ru-G_4OH PAMAM dendrimer nanocomposite precursor and reduced for 2 h at 300°C [72].

though still well-dispersed on the alumina support surface. The increased in particle sizes was attributed to sintering which may have occurred during the dendrimer removal process. No catalytic applications experiments were reported for these catalysts.

The same research group investigated the effect of support material on the sintering by immobilisation of dendrimer-derived Ru nanoparticles onto three different support materials (low-surface-area γ-alumina, high-surface-area γ-alumina, and SiO_2) few years later [73]. In this case, the ability of dendrimers to allow for the synthesis of particles with varying sizes was tested by using different generations (G 4–5) of hydroxyl-terminated PAMAM dendrimers as well as dendrimer-to-metal molar ratios were used to prepare the Ru-DEN precursors. For instance, for generation 4 dendrimers, a dendrimer-to-metal molar ratio of 1:20 and 1:40 were used, while for generation 5, dendrimer to metal ion molar ratios of 1:60 and 1:100 were used. The average particle sizes determined by high resolution TEM were measured to be 0.9 nm and 1.2 nm for 1:20 and 1:40 molar ratios, respectively. For generation 5, particle sizes for 1:60 and 1:100 molar ratios were determined to both have average sizes of 1.4 nm. These average particle sizes were found to be 10–20% greater than the expected calculated theoretical average sizes. As in the previous case, the immobilisation of these DEN precursors onto different support materials resulted in an increase of particle sizes by 75%, attributed to sintering. However, this sintering was found not to be induced by the type of support materials used.

4.1 Synthesis, characterisation and catalytic applications of unsupported colloidal mono- and bimetallic Ru-DENs

4.1.1 Monometallic Ru-DENs/DSNs for reduction of 4-Nitrophenol

4-Nitrophenol (4-NP) and its derivatives are considered among some of the major pollutants, hazardous waste [74]. The United State Environmental Protection Agency (US EPA) has labelled these pollutants as priority toxic pollutants [1]. 4-NP and its derivatives are generally generated from agricultural sectors and manufacturing industries. For example, it is a waste product from companies that manufacture explosives, dyes, and other chemical products [75]. This has rendered these organic compounds to be the most common non-biodegradable and toxic persistent pollutants in industrial, agricultural wastewater and aquatic environment [76]. Moreover, 4-NP is produced in large quantities worldwide and is considered the most toxic as compared to other mono-nitrophenols [1]. The presence of nitrogen group on this compound makes it more stable, as a consequence, chemical and biological degradation of wastewater polluted with 4-NP is challenging. Therefore, the development of sustainable and environmentally friendly methods to remove these compounds from wastewater is of great importance.

More recently, Ru-DEN catalysts have been found to exhibit excellent activity in the conversion of 4-NP to 4-aminophenol (4-AMP) and other organic pollutants in the presence of excess reducing agents such as $NaBH_4$ or H_2 [38]. 4-AMP, on the other hand, is a commercially valuable intermediate for the synthesis of anti-pyretic and analgesic drugs [77]. The reduction of 4-NP is analysed by monitoring the change in the absorbance peak at around λ 400 nm (maximum absorption peak for 4-NP) and the emergence of a new absorption peak around λ 310 nm, which is attributed to the formation of the 4-AMP product. This reaction is accompanied by the formation of isosbestic points on the UV-vis spectra, signalling that there is no side product (except 4-AMP) formed.

One of the earliest example of Ru-DENs catalysed reduction of 4-NP to 4-AMP can be traced back to the work published by Antonels and Meijboom [78]. Generation

4–6 of commercial hydroxyl-terminated PAMAM dendrimers were employed for the synthesis of Ru-DENs with varying particle sizes. Characterisation techniques such as IR, UV-vis, and TEM were used for the synthesis process as well as determining the resulting Ru-DENs morphology and sizes. Notably, the increase in dendrimer generation resulted in an increase in average particle sizes. For instance, the nanoparticles, with almost-spherical shape, prepared using generation 4, 5, and 6 PAMAM dendrimers were determined to have average particle sizes of 1.2 ± 0.07 nm, 1.4 ± 0.12 nm, and 2.2 ± 0.26 nm, respectively. It should be noted, however, that different metals to dendrimer molar ratios were used for different dendrimer generation. **Figure 4** shows TEM images and particle size distribution reported in Ref. [78].

The catalytic investigation of these Ru-DENs catalysts was performed in the selective reduction of 4-NP to produce 4-AMP. This reaction was set up under first-order

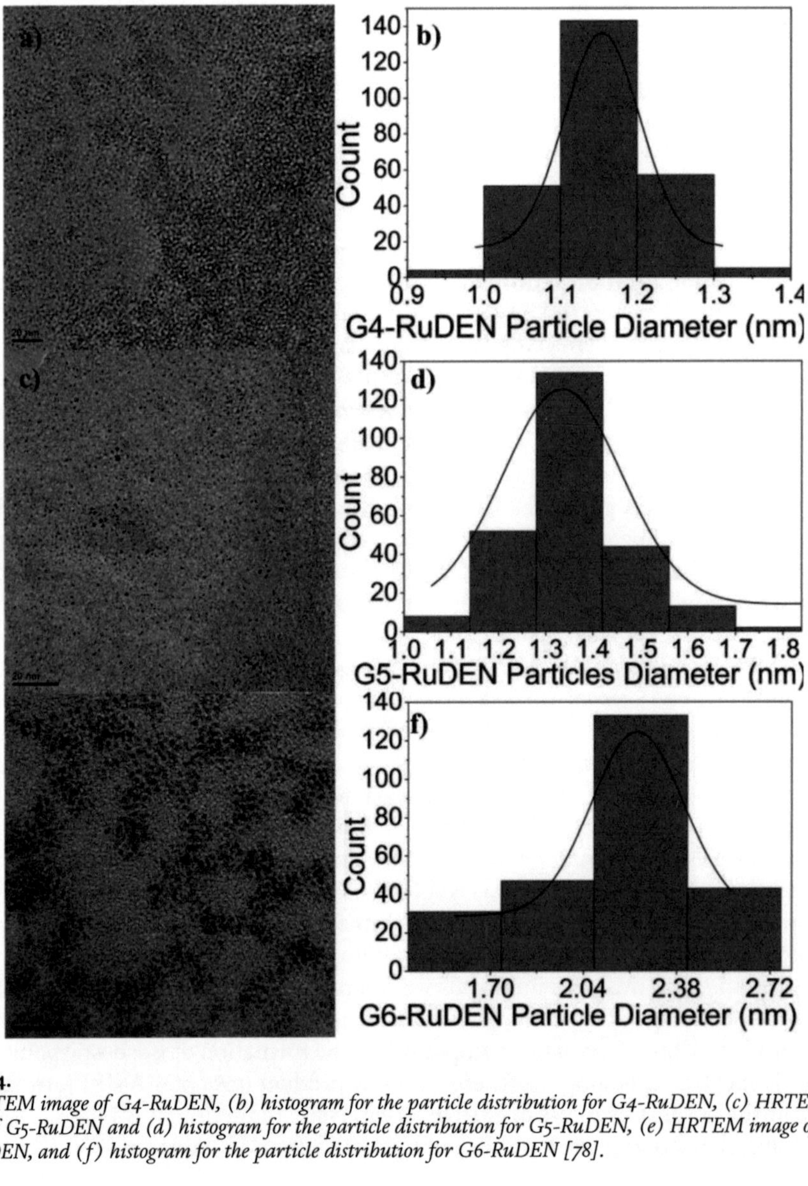

Figure 4.
(a) HRTEM image of G4-RuDEN, (b) histogram for the particle distribution for G4-RuDEN, (c) HRTEM image of G5-RuDEN and (d) histogram for the particle distribution for G5-RuDEN, (e) HRTEM image of G6-RuDEN, and (f) histogram for the particle distribution for G6-RuDEN [78].

conditions, by ensuring that the reducing agent, $NaBH_4$, is always in large excess relative to the 4-NP substrate. The calculated apparent rate constants (k_{app}) for the generation 4, 5, and 6 dendrimer catalysts were determined to be $3.96 \times 10^{-4}\ s^{-1}$, $6.40 \times 10^{-4}\ s^{-1}$, and $2.90 \times 10^{-4}\ s^{-1}$, respectively. The kinetic analysis was done based on the assumption that the reaction occur on the nanoparticle surface and hence follow the Langmuir-Hinshelwood mechanism. For this purpose, two sets of kinetic data were collected. One set of k_{app} data was collected by varying the concentration of 4-NP while keeping BH_4^- concentration and temperature constant. The second set of k_{app} data was collected by varying the BH_4^- concentration while the 4-NP concentration and temperature were kept constant. It was observed that the apparent rate constant appear to increase with the increase in BH_4^- concentration. On the contrary, the k_{app} was found to decrease as the concentration of 4-NP is increased. The observed decrease in the apparent rate constant as 4-NP is increased was attributed to the fact that as the more 4-NP is injected into the reaction system, the catalyst's surface becomes more saturated as less active atoms are available for the BH_4^- to react with the catalyst. This suggests that there is a competition for adsorption between the substrate and the reactant with the nanoparticle surface.

Another example of Ru-DENs catalysed reduction of 4-NP was reported by Murugan and Pakrudheen [79]. Their study entails the initial synthesis of the generation 2 amphiphilic poly(propylene imine) (PPI) dendrimer template. The characterisation of the synthesised dendrimer template was done using various techniques such as FT-IR, NMR, and MALD-TOF. The as-synthesised dendrimer was further utilised for the stabilisation and encapsulation of Ru nanoparticles (Ru-DENs), which were characterised using UV-vis, FT-IR, TEM, and XRD. For example, characterisation by TEM revealed that the formed Ru-DENs had an average particle size of 0.8 ± 0.3 nm. The formation of these ultra small-sized Ru-DENs was attributed to the smaller hydrodynamic diameter of PPI dendrimers as compared to that of PAMAM counterparts. Therefore, the interaction between the metal ions and the tertiary amine of the PPI dendrimers is believed to be stronger, allowing for the formation of smaller nanoparticles. The catalytic properties of the prepared Ru-DENs were evaluated in the reduction of 4-NP to yield 4-AMP. As with other related studies, the pseudo-first-order reaction kinetics conditions were applied in this case too. The observed reaction rate constant of $15.95 \times 10^{-4}\ s^{-1}$. The increase in the concentrations of BH_4^- and the catalyst while keeping other parameters constant was found to result in an increase in the reaction rate constant. For instance, an increase in the borohydride concentration from 20 mM to 60 mM resulted in the increase of the rate constant from $4.66 \times 10^{-4}\ s^{-1}$ to $8.92 \times 10^{-4}\ s^{-1}$. Similarly, an increase in the rate constant from 7.93 to 15.95 was observed as the concentration of the catalyst is varied from 0.02 mM to 0.08 mM.

4.1.2 Bimetallic of Ru-based DENs/DSNs catalysts for reduction of 4-Nitrophenol and hydrogenation reactions

A synthesis of a series of Ru-Ni bimetallic DEN catalysts via intradendrimer redox displacement was reported by Marvin et al. [80]. In their study, nickel was chosen as an appropriate metal partner for a displacement method to prepare Ru and RuNi bimetallic DEN catalysts. Generally, the displacement synthetic method was carried out by the initial formation of Ni DENs (1:30 of dendrimer: metal ion molar ratio) using commercially obtain generation 4 hydroxyl-terminated PAMAM dendrimer. The synthesis of Ru or bimetallic RuNi DENs can be obtained by addition of equimolar of Ru^{3+} or any molar ratio smaller than that initially used for Ni^{2+} for the formation of Ni DENs, respectively. For instance, a variety of bimetallic RuNi DEN catalysts were prepared via displacement using molar excess ratios ranging between 5 and 20 for Ru^{3+} and Ni^{2+}. The resulting bimetallic DEN catalysts were labelled as Ru_5Ni_{30}, $Ru_{10}Ni_{30}$, $Ru_{15}Ni_{30}$, and $Ru_{20}Ni_{30}$-DENs. The catalytic activity of these DEN catalysts was evaluated in the reduction of 4-NP. All RuNi bimetallic DEN catalysts were found to display a superior activity as compared to either monometallic Ru- or Ni-DEN catalysts. Among all bimetallic DEN catalysts, the highest average rate constant (2.2 s^{-1}) was observed for the $Ru_{10}Ni_{30}$ DEN catalyst. On the contrary, the nickel dominated bimetallic catalyst (Ru_5Ni_{30}) showed the smallest average catalytic rate (0.95 s^{-1}).

Wang and Peng reported the synthesis of new class of fourth-generation poly(propylene imine) dendrimer functionalised with 32 triolefinic 15-membered microcycles on the surface (G5-M). The synthesised G4-M dendrimer was characterised using NMR, IR, and elemental analysis (EA). The prepared G4-M dendrimer was utilised for the fabrication and stabilisation of bimetallic RuRh nanoparticles (RuRh-DSNs) via co-complexation method [81]. Stabilisation was achieved by ensuring that metal ions of Ru and Rh are coordinated with both the surface groups and the tertiary of the PPI dendrimer prior to the reduction of the dendrimer-metal ions composite with excess $NaBH_4$. In this study, dendrimer stabilised bimetallic RuRh nanoparticles with various molar ratios (G4-M (Ru_xRh_y)) were prepared and characterised using high-resolution TEM and EDS techniques to determine the particle sizes and morphology. The average nanoparticles diameter for the G4-M (Ru_7Rh_3), G4-M (Ru_5Rh_5), G4-M (Ru_3Rh_7) bimetallic catalysts were determined to be 6.1 ± 1.4 nm, 4.5 ± 1.5 nm, 7.7 ± 2.1 nm, respectively. The morphology of these nanoparticles was observed to be nearly uniform and almost spherical with a Gaussian-like size distribution.

The catalytic activity of the RuRh bimetallic nanoparticles was investigated in the hydrogenation of nitrile-butadiene rubber (NBR). The hydrogenation capability of these catalysts was measured in terms of the degree of hydrogenation (HD). All bimetallic RuRh DSNs were found to give larger HD as compared to the corresponding monometallic Ru or Rh DSNs, despite the latter having smaller average particle sizes. The G4-M (Ru_3Rh_7 catalyst was found to exhibit higher catalytic hydrogenation activity than the other evaluated two bimetallic ones. The recycling study performed using the G4-M (Ru_3Rh_7 catalyst showed that only a slight loss of catalytic activity was observed even after 3 reaction cycles. Higher catalytic activities displayed by the bimetallic catalysts were attributed to the electronic effect. Some other catalytic reactions involving monometallic and bimetallic colloidal Ru-based DEN catalysts have been summarised in **Table 1**.

4.2 Synthesis, characterisation and catalytic applications of supported mono- and bimetallic Ru-DENs

These coming sections will focus on the synthesis and characterisation of supported monometallic Ru-DEN and bimetallic Ru-based DEN catalysts as well as their catalytic applications in the hydrogenation reactions involving aromatic compounds (such as phenols), methanol and biomass-derived molecules (such as levulinic acid).

4.2.1 Monometallic supported Ru-DENs/DSNs catalysts for hydrogenation reactions

More recently, an interesting study on the synthesis of new two types of second-generation dendrimers using the divergent approach has been reported by Keshtiara et al. [91]. The first-generation dendrimer (a precursor for second-generation) was

Catalyst	Substrate	Time (h)	Conversion (%)	Current density/ mA•¯cm^2	TOF (h^{-1})	Ref
Ru_{40}-DENs	LA	5	>99		5025	[82]
Ru_{40}@Meso-TiO_2	LA	5	92		3878	[82]
Ru_{40}@Meso-SiO_2	LA	5	94		5314	[82]
G4-RuSil60	Citral	4	84		—	[83]
G5-RuSil60	Citral	4	22		—	[83]
G6-RuSil60	Citral	4	12		—	[83]
0.15 wt% G4-RuSil100	Toulene	—	73.5		—	[84]
Ru-DENs	PMHS	2	44		—	[85]
Ru7Rh3-DENs	PMHS	2	71		—	[85]
Ru5Rh5-DENs	PMHS	2	79		—	[85]
G4OH-$Pt_{20}Ru_{20}$/C[a]	Methanol	—	—	17.8 ± 2.4	—	[86]
G4OH-$Pt_{20}Ru_{20}$/C[b]	Methanol	—	—	13.6 ± 4.6	—	[86]
20% Pt-Ru PAMAM/ CNF	Methanol	—	—	0.25	—	[87]
PPI-dendr-PEG-Ru-90	Benzene	2	99		—	[88]
PPI-dendr-PEG-Ru-20	Benzene	2	>85		—	[88]
PPI-dendr-PEG-Ru-90	Phenol	2	>80		—	[88]
PPI-dendr-PEG-Ru-20	Phenol	2	>70		—	[88]
Ru_{np}-nSTDP	Cyclooctene	4	93		13.84	[89]
Ru_{np}-nSTDP	Cyclohexene	4	90		13.39	[89]
Ru_{np}-nSTDP	Benzyl alcohol	4	90		17.85	[89]
Dendrimer-Derived PtRu/SiO_2	3,4-epoxy-1-butene	—	—		1162[c]	[90]

PMHS = poly(methylhydro)siloxane.[a]Catalyst activated at 500°C.
[b]Catalyst activated at 400°C.
[c]min^{-1}.

Table 1.
Summary of dendrimer stabilised/encapsulated Ru-based catalysed reactions.

prepared from hexakis(bromomethyl)benzene (hex) core and vanillin branches. The second step for the synthesis of two types of second-generation dendrimers involves the reaction of the first-generation dendrimer precursor with *ortho*-aminopyridine (py) and *para*-aminobenzoic acid (bza) to produce second-generation dendrimers, hexpyD and hexbzaD, respectively. The produced dendrimers were characterised using techniques such as attenuated total reflectance-infrared (ATR-IR), field-emission scanning electron microscopy (FE-SEM), NMR, and TEM, among others. These dendrimers were subsequently reacted with Ru^{3+} ions (in DMF solvent) to form hexanuclear Ru (III) chloride complexes. The formed complexes were each immobilised on a silica support in 2-methoxyethanol solvent and allowed to mix for 24 h. An excess of $NaBH_4$ was then added to these complexes-silica mixtures, resulting in the formation of dendrimer encapsulated/stabilised Ru nanoparticles immobilised on silica support (S-D/RuNPs).

TEM analysis of the S-D/TEM catalysts revealed that the Ru nanoparticles synthesised from hexpyD and hexbzaD have average sizes ranging from 0.2–4 nm to 0.6–20 nm, respectively. Small sized-Ru nanoparticles obtained for S-D/RuNPs composite material was attributed to the steric hindrance that exists on the periphery of the hexpyD dendrimer template. **Figure 5** shows TEM images (captured on different scales) for silica supported Ru nanoparticles synthesised using hexpyD dendriemer. These TEM results correlated well with the XRD analysis. For example, the broad peaks observed around $2\theta = 21.8°$ on the XRD patterns for S-D/RuNPs synthesised from hexpyD and hexbzaD (**Figure 6a** and **b**) were attributed to the presence of amorphous silica. While the peaks observed at $2\theta = 21.8°$ (**Figure 5a** and **b**) were assigned to the existence of RuO that may have formed because of the surface oxidation of metallic Ru NPs. On the other hand, isotherms collected from BET N_2 adsorption-desorption analysis showed characteristics of mesoporous materials (type IV isotherms) and possess relatively high BET surface area and narrow pore size distribution (see **Figure 7a** and **b**). The specific surface area of 165 and 303 m^2/g was measured for S-D/RuNPs derived from hexpyD and hexbzaD, respectively.

The catalytic evaluation of these S-D/RuNPs catalysts was performed in the hydrogenation of citral to 3,7-dimethyloctanol. Significant conversion rates (about 40%) were observed for both catalyst only 30 min of the reaction at the optimum reaction conditions applied and as such, turnover frequencies (TOFs) for these catalysts were calculated based on the first 30 min of the reaction. The S-D/RuNPs prepared using

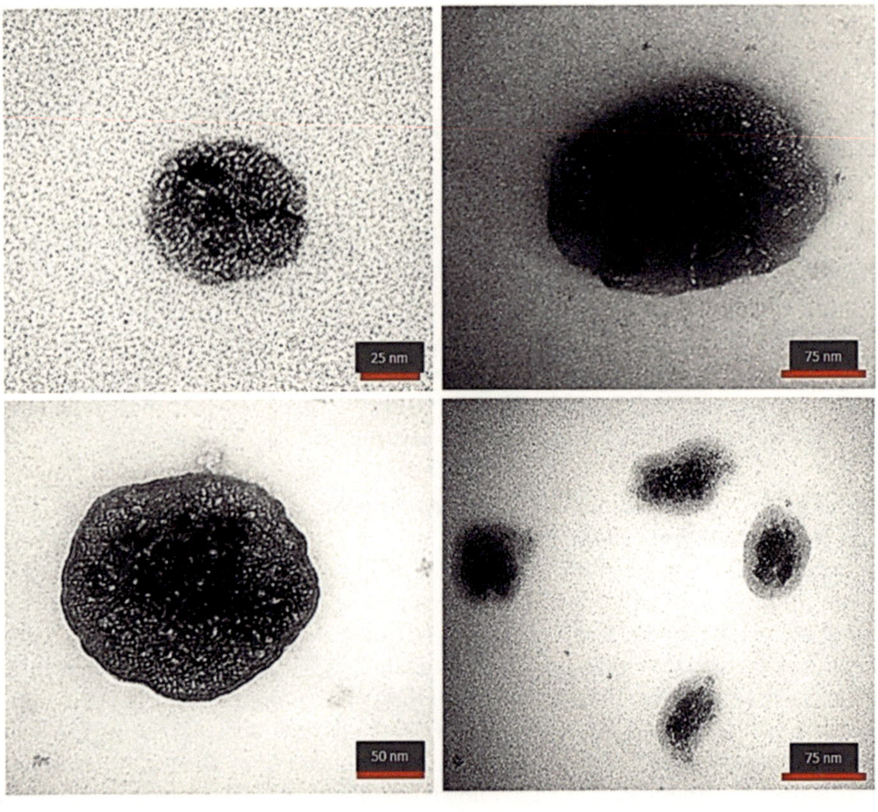

Figure 5.
TEM images of S-D/RuNPs prepared from hexpyD [91].

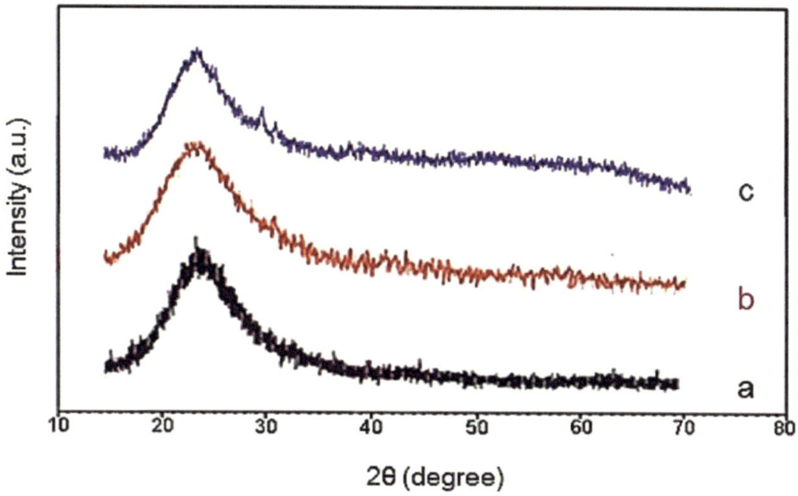

Figure 6.
XRD patterns of amorphous silica (a), and S-D/RuNPs prepared from hexpyD (b) and hexbzaD (c) [91].

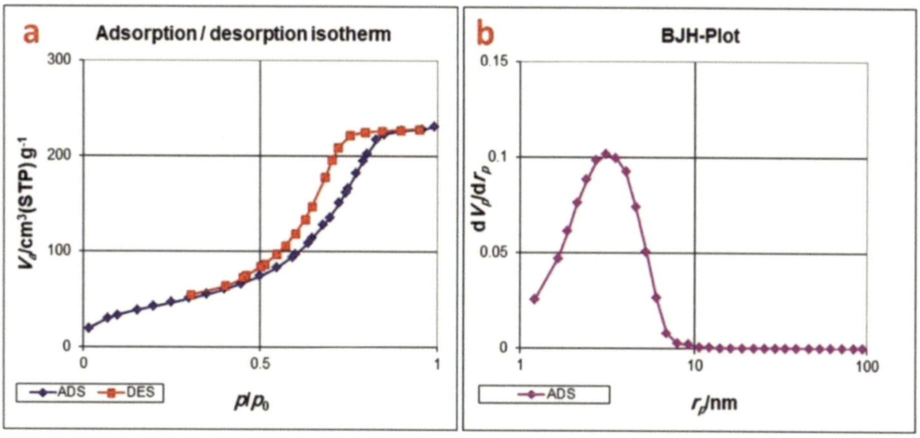

Figure 7.
Nitrogen adsorption-desorption isotherms (77 K) (a) and pore size distribution (b) of S-D/RuNPs prepared from hexpyD [91].

the hexbzaD dendrimer was found to exhibit greater TOF as compared to the one prepared using the hexpyD dendrimer. This was attributed to less sterically hindered surfaces of the hexbzaD dendrimer as well as the large particles for RuNPs prepared from hexbzaD, which provide a large number of adsorption sites.

The hydrogenation of alkyl-substituted phenols to the corresponding alkyl substituted cyclohexanols over organo-silica supported dendrimer-stabilised Ru nanocatalysts catalyst has also been reported by Karakhanov et al. [92]. In this case, two types of silica support materials were fabricated through the modification of the dendrimer template. Ideally, the PPI dendrimers were modified with (3-glyidoxy-propyl)trimethoxysilane using ethanol as a solvent. After this, the acid-catalysed hydrolysis and polymerisation of $Si(OEt)_4$ occur for the formation of ordered porous dendrimer-silica materials (dendr-G3-SiO_2). Alternatively, mesoporous dendrimer-silica materials were prepared by the addition of aqueous Pluronic P123 polymer to the previously modified dendrimer-silica materials (dend-G3-*meso*-SiO_2). The PPI dendrimers were playing the templating role for the pore formation for both cases. These materials were thoroughly characterised using IR, XPS, NMR, and TEM to confirm the coordination of the dendrimer with the silica support.

The synthesis of Ru nanoparticles immobilised onto these dendrimer-modified silica supports was carried out by using the wetness impregnation method, which involved the addition of an aqueous solution of Ru^{3+} ions to either of the dendr-G3-SiO_2 or dendr-G2-*meso*-SiO_2 support networks. This resulted in the coordination of Ru ions with the amino groups of the dendrimer frameworks. The reduction of Ru^{3+} ions was carried out using $NaBH_4$ in water-ethanol medium, at room temperature, resulting in the formation of dendr-G3-SiO_2-Ru and dendr-G2-*meso*-SiO_2-Ru catalysts. The average Ru nanoparticles in the dendr-G3-SiO_2-Ru and G2-*meso*-SiO_2-Ru catalysts, as determined byTEM analysis were found to be 0.97 ± 0.09 nm and 1.28 ± 0.09 nm, respectively. These catalysts were evaluated in the hydrogenation of alkyl substituted phenols. The dendr-G2-*meso*-SiO_2-Ru catalyst was found to generally exhibit superior activity as compared to the dendr-G3-SiO_2-Ru catalyst. For instance, the hydrogenation of meta-substituted phenol turnover frequency for the hydrogenation of meta-substituted phenol was calculated to be 6090 h^{-1} for the dendr-G2-*meso*-SiO_2-Ru catalysts as compared to 1841 h^{-1} obtained for the dendr-G3-SiO_2-Ru catalyst under similar reaction conditions. Both catalysts, however, displayed appreciable stability as could be re-used several times without any significant loss in their initial activities.

The same research group of Karakhanov also reported the hydrogenation of a sustainable, biomass-derived molecule, levulinic acid (LA) and its esters to γ-valerolactone (GVL) catalysed by similar type of catalysts (G3-dendr-SiO_2-Ru G2-dendr-*meso*-SiO_2) as those used for the hydrogenation of phenols [93]. The reaction rate and the yield for GVL were found to be influenced by several factors such as temperature, H_2 pressure, solvent, and support nature. For example, the mesoporous silica supported dendrimer stabilised Ru catalyst gave quantitative GVL yields (above 95%) within 2 h when the reaction temperature is set at 120°C. However, when the reaction temperature is decreased from 120 to 80°C, it took 6 h to obtain a quantitative yields of GVL. Similarly, when excess amount of water solvent is used (LA:H_2O = 1:2) the conversion of LA was found to also increase. The promoting effect of water on the hydrogenation of LA has also been reported by other authors [71]. Just like in the case of catalytic hydrogenation of phenols (using similar catalysts) reported by this group, the G2-dendr-*meso*-SiO_2 catalyst was found to display superior activity compared to its counterpart, G3-dendr-SiO_2-Ru catalyst under similar reaction conditions. This enhanced activity observed for the G2-dendr-*meso*-SiO_2 catalyst was attributed to the carrier structure of the catalyst.

4.2.2 Bimetallic supported Ru-DENs/DSNs catalysts for hydrogenation and oxidation reactions

Bimetallic Ru-containing DEN catalysts have also been reported to exhibit superior catalytic activity, particularly in reactions such as oxidation of methanol. Of these Ru-based bimetallic DEN catalysts, supported Pt-Ru dendrimer encapsulated/stabilised nanoparticles appear to have been well investigated, especially in the oxidation of methanol. For example, Calderón et al. reported the electro-oxidation of methanol catalysed by carbon-nanofibers (CNF) supported dendrimer stabilised Pt-Ru nanoparticles [94]. Different generations (G0–3) of PAMAM dendrimers were used for the stabilisation of the synthesised supported Pt-Ru bimetallic nanoparticles in this case. The dendrimer generation was found to have an effect on the resulting bimetallic nanoparticles. For instance, TEM analysis revealed that nanoparticles synthesised using generation zero had low dispersion on the CNF support. Moreover, the average particle size was found to decrease with an increase in dendrimer generation. For example, the average nanoparticle size of 3.7 ± 0.1 nm was obtained for generation-zero dendrimer, while generation three nanoparticles had an average size of 1.9 ± 0.4 nm prior to heat treatment. The heat treatment applied to clean the surface of the nanoparticles (and possibly remove the organic dendrimer template) also generally resulted in an increase in average particle sizes irrespective of the dendrimer generation.

Preliminary catalytic activity of both sets of catalysts (non-heat treated and heat-treated) in the electrochemical oxidation of methanol was evaluated in the supporting electrolyte. The non-heat treated catalysts, particularly those prepared in generation 0, 2, and 3 dendrimers, showed high capacitive currents for the hydrogen adsorption-desorption process. The decrease in capacitive currents was observed for all heat-treated catalysts, indicating a possible removal of the dendrimer template during heat treatment process of the catalysts. The catalytic activities for electrochemical oxidation of methanol for all catalysts were done using voltamperometric studies. All heat-treated catalysts were found to have higher activities (high current densities) for methanol oxidation. This was attributed to the high content of Ru oxides on the surface of these heat treated catalysts, which is believed to enhance the oxidation of methanol.

Bimetallic PdRu nanoparticles prepared based on the stabilisation using third-generation PPI dendrimers crosslinked by hexamethylene diisocynate (HMDI) support have also been utilised for the catalytic hydrogenation of aromatic compounds such as benzene and its derivatives under two-phase condition [95]. The synthesised bimetallic PdRu nanoparticles were determined to have a small average size of 0.8 ± 0.1 nm by TEM Although the bimetallic PdRu catalyst showed an increased hydrogenation rate for benzene substrate (due to synergistic effect of Pd and Ru) as compared to its monometallic Ru catalyst counterpart, its (bimetallic) activity drastically dropped in the case of substituted aromatic compounds. This was attributed to the steric hindrance that exist in the third-generation dendrimer structure. **Table 1** contains a summary of the reported catalytic system involving supported monometallic and bimetallic dendrimer-derived Ru catalysts.

5. Effect of optimization/modification of dendrimer macromolecules on the size of the NPs

Cognizant of all the discussion on the leveraging of dendrimers for stabilisation and encapsulation of metal nanoparticles outlined in this chapter, it is imperative to highlight that the optimization of the dendrimer macromolecule structure has also been found to have an effect on the size of the formed NPs. The dendrimer macromolecules condition can be can be optimised in few ways. For instance, peripheral groups of the dendrimer can be functionalized or protonated. For example, dendrimer-metal composite synthesised using amine- or carboxylic acid-terminated dendrimers usually favours the complexation of the metal ions on both the periphery groups and the tertiary amines. Reduction of such dendrimer-metal ions composite will yield the mixture of both inter- and intradendrimer nanoparticles. The interdendrimer nanoparticles (DSNs) are usually bigger than the intradendrimer nanoparticles (DENs). This is because DSNs are stabilised by primary groups and are therefore prone to agglomeration as opposed to the DENs. In order to avoid the formation of a mixture of both DSNs and DENs, the peripheral groups can be functionalized or protonated (pH adjustment) [42]. For example, the pH of the aqueous dendrimer solution can be adjusted to be in the range between pH 2 and 5 to protonate the primary amine groups, leaving tertiary available for complexation with metal ions. Subsequent reduction will favour the formation of small-sized, monodispersed, and nearly spherical DENs. Alternatively, the surface groups can be functionalized to quaternary ammonium terminated groups to circumvent complexation of metal ions with primary groups, which in turn promote the formation of DENs.

6. Conclusions

Dendrimers can be leveraged for the synthesis of stable and catalytically active Ru (and nanoparticles (and other noble metals). Different nanoparticle sizes of the Ru-DEN catalysts can be tailored by manipulation of the dendrimers-to-metal ions molar ratios. The size of the dendrimer and the encapsulated nanocatalyst can have an effect on the catalytic performance. Although the synthesis of these Ru-DENs (and other metal DENs) may look straightforward in theory, it is actually tedious and has to be carried out cautiously to avoid errors as very low concentrations are used, particularly for academic research purposes. Dendrimer-derived Ru catalysts have been proven to be excellent catalysts in the hydrogenation and oxidation reactions involving several substrates such as phenols, citral, and methanol. The use of emerging technologies such as liquid handling robotics can also assist in minimising these potential human errors. Although there is a plethora of research publications for the synthesis of DEN catalysts emerging from academic researchers around the globe, it must poignantly be hinted that commercial dendrimers are excessively commercially expensive for emerging researchers to afford, particularly from developing countries as those found in the African continent. Therefore, research on the use of low-cost, “in-house” synthesised dendrimers for nanocatalysts encapsulation must be promoted to enhance participation from emerging young researchers coming from poor countries or institutions. Additionally, although the use of these types of catalysts (DENs) have shown great potential to catalyse various reactions, their use in the manufacturing industries is still lacking.

Dendrimer-derived bimetallic Ru-based catalyst proved to have superior activity compared to monometallic Ru catalysts, particularly in reactions such as methanol oxidation and hydrogenation of aromatic compounds. However, the synthesis of Ru-DENs catalysts using sustainable and environmentally friendly reagents is yet to be exploited. For example, the use of environmentally friendly precursors and solvents for the synthesis of metal catalysts is still to be fully exploited. More importantly, synthetic methods that make use of sustainable and environmentally friendly templating agents such as biopolymers (such as chitosan) for the encapsulation of nanocatalysts must be developed.

Author details

Mulisa Maumela[1]* and Ndzondelelo Bingwa[2]

1 Department of Mathematics, Science and Technology Education, University of Limpopo, Sovenga, South Africa

2 Center for Synthesis and Catalysis, Department of Chemical Sciences, University of Johannesburg, Johannesburg, South Africa

*Address all correspondence to: mulisa.maumela@ul.ac.za

References

[1] 4-Nitrophenols. Washington, DC: Health and Environmental Effects Profile in US Environmental Protection Agency; 1980. p. 135

[2] Hu H, Xin JH, Hu H, Wang X, Miao D, Liu Y. Synthesis and stabilization of metal nanocatalysts for reduction reactions—A review. Journal of Materials Chemistry A. 2015;**3**:11157-11182

[3] Yang Q, Xu Q, Jiang HL. Metal-organic frameworks meet nanoparticles: Synergistic effect for enhanced catalysis. Chemical Society Reviews. 2017;**46**:4774-4808

[4] Hunt ST, Milina M, Alba-Rubio AC, Hendon CH, Dumesic JA, Romáin-Leshkov Y. Self-assembly of Noble metal monolayers on transition metal carbide nanoparticle catalysts. Science. 2016;**352**:974-978

[5] Saxena V, Kumar N, Saxena VK. A comprehensive review on combustion and stability aspects of metal Nanoaprticles and its additive effect on diesel and biodiesel fuelled C.I. Engine. Renewable and Sustainable Energy Reviews. 2017;**70**:563-588

[6] Hvolbaek H, Janssens TVW, Bjerne BS, Falsig H, Christensen CH, Nørskov JK. Catalytic activity of Au nanoparticles. NanoToday. 2007;**2**:14-18

[7] Narayan N, Meiyazhagan A, Vajtai R. Metal nanoparticles as green catalysts. Materials. 2019;**12**:1-12

[8] Majedi SM, Lee HK. Recent advances in the seperation and quantificatrion of metallic nanoparticles and ions in the environment. TrAC Trends in Analytical Chemistry. 2016;**75**:183-196

[9] Shylesh S, Schünemann V, Thiel WR. Magnetically seperable nanocatalysts: Bridges between homogeneous and heterogenenous catalysis. Angewandte Chemie, International Edition. 2010;**49**:3428-3459

[10] Chechik V, Crooks RM. Dendrimer-encapsulated Pd nanoparticles as fluorous phase-soluble catalysts. Journal of the American Chemical Society. 2000;**122**:1243-1244

[11] Nemanashi M, Meijboom R. "Cat in bag" recycling of denderimer encapsulated Au nanoparticles by dialysis membrane in the reduction of 4-nitrophenol: Proof of heterogeneous catalysis. Catalysis Communications. 2016;**83**:53-57

[12] Astruc D, Lu F, Aranzaes JR. Nanoparticlesc as recyclabe catalysts: The frontier between homogeneous and heterogeneous catalysis. Angewandte Chemie, International Edition. 2005;**44**:7852-7872

[13] Tabor C, Narayanan R, El-Sayed MA. Catalysis with transition metal nanoparticles in colloidal solutions: Heterogeneous or homogeneous? Model Systems in Catalysis. 2009:395-414

[14] Jamkhande PG, Ghule NW, Bamer AH, Kalaskar MG. Metal nanoparticles synthesis: An overview on methods of preparation, advantages and dissadvantages, and applications. Journal of Drug Delivery Science and Technology. 2019;**53**:1-11

[15] Meyers MA, Mishra A, Benson DJ. Mechanical properties of nanocrystalline materials. Progress in Materials Science. 2006;**51**:427-556

[16] Mukherjee P, Ahmad A, Mandal D, Sainkar SR, Khan MI, Parishcha R, et al. Fungus mediated synthesis of silver nanoparticles and their immobilization in the mycelia matrix: A novel biological approach to nanoparticle synthesis. Nano Letters. 2001;**1**:515-519

[17] Gröhn AJ, Prastinis SE, Sánchez-Ferrer A, Mezzenga R, Wegner K. Scale-up of nanoparticle synthesis by flame spray pyrolysis: The high-temperature particle residence time. Industrial and Engineering Chemistry Research. 2014;**53**:10734-10742

[18] Kimling J, Maier M, Okenve B, Kotaidis B, Ballot V, Plech H. Turkevich method for gold nanoaprticle synythesis revisited. The Journal of Physical Chemistry. B. 2006;**110**:15700-15707

[19] Karakhanov EA, Maskimov AL, Zolotukhina AV, Kardasheva YS. Hydrogenation catalysts based on metal nanoaprticles stabilized by organic ligand. Russian Chemical Bulletin. 2013;**5**:1465-1492

[20] Gonzalez-Galvez D, Lara P, Rivada-Wheelaghan O, Conejero S, Chaudret B, Philippot K, et al. NHC-stabilized ruthenium nanoaprticles as new catalysts for the hydrogenation of aromatics. Catalysis Science & Technology. 2013;**3**:99-105

[21] Lara P, Philippot K, Chaudret B. Organometallic ruthenium nanoaprticles: A comparative study of the influence of the stabilizers on their characteristics and reactivity. ChemCatChem. 2013;**5**:28-45

[22] Maskimov AL, Kuklin SN, Kardasheva YS, Karakhanov EA. Hydrogenation of phenols in ionic liquids on rhodium nanoparticles. Petroleum Chemistry. 2013;**53**:157-163

[23] Dawson R, Cooper AL, Adams DJ. Nanoporous organic polymer networks. Progress in Polymer Science. 2012;**37**:530-563

[24] Axet MR, Philippot K. Catalysis with colloidal ruthenium nanoparticles. Chemical Reviews. 2020;**120**:1085-1145

[25] Hasegawa S, Tsukuda T. Exploring novel catalysis using polymer-stabilized metal clusters. Bulletin of the Chemical Society of Japan. 2021;**94**:1036-1044

[26] Kuklin SN, Maximov A, Zolotukhina AV, Karakhanov EA. New approach for highly selective hydrogenation of phenol to cyclohexanone: Combination of rhodium nanoparticles and cyclodextrins. Catalysis Communications. 2016;**73**:63-68

[27] Noël S, Léger B, Ponchel A, Philippot K, Denicourt-Nowicki A, Roucoux A, et al. Cyclodextrin-based systems for the stabilization of metallic (0) nanoaprticles and their versatile applications in catalysis. Catalysis Today. 2014;**235**:20-32

[28] Liu Y, Hartman RL. Reaction kinetics of a water-soluble palladium-β-cyclodextrin catalyst for a suzuki-miyaura cross-coupling in continous flow. Reaction Chemistry and Engineering. 2019;**4**:1341-1346

[29] Cocq A, Bricout H, Djedaïni-Pilard F, Tilloy S, Monflier E. Rhodium-catalyzed aqueous biphasic olefin hydroformylation promoted by amphiphilic cyclodextrins. Catalysts. 2020;**10**:56

[30] Huc V, Pelzer K. A new specifically designed calix[8]arene for the synthesis of functionalized, nanometric and subnanometric Pd, Pt, and Ru nanoparticles. Journal of Colloid and Interface Science. 2008;**318**:1-4

[31] Zhang Q, Catti L, Tiefenbancher K. Catalysis inside the hexameric

resorcinarene capsule. Accounts of Chemical Research. 2018;**51**:2107-2114

[32] Gaeta C, Talotta C, Rosa MD, Manna PL, Soriente A, Neri P. The hexameric resorcinarene capsule at work: Supramolecular catalysis in confined spaces. Chemistry - A European Journal. 2019;**25**:4899-4913

[33] Zhang W, Sun Y, Zhang L. Fabrication of high efficient silver nanoparticles catalyst supported on poly(gylcidylmethacrylate)-polyacrylamine. Industrial and Engineering Chemistry Research. 2016;**55**:12398-12406

[34] Malakhova I, Privar Y, Parotkina Y, Mironenko A, Elisseikina M, Balatskiy D, et al. Rational design of polyamine-based cryogels for metal ion sorption. Molecules. 2020;**25**:4801

[35] Dzhardimalieva GI, Uflyand IE. Preparation of metal-polymer nanocomposites by chemical reduction of metal ions: Functions of polymer matrices. Journal of Polymer Research. 2018;**25**:255

[36] Ohde H, Wai CM, Kim H, Kim J, Ohde M. Hydrogenation of olefins in supercritical CO_2 catalyzed by palladium nanoparticles in a water-CO_2 microemulsion. Journal of the American Chemical Society. 2002:4540-4541

[37] Yoon B, Kim H, Wai CM. Dispersing palladium nanoparticles using a water-in-oil microemulsion-homogenization of heterogeneous catalyisis. Chemical Communications. 2003:1040-1041

[38] Illunga AK, Meijboom R. A review of dendrimer-encapsulated metal nanocatalysts applied in the fine chemicals transformation. Catalysis Letters. 2019;**159**:84-99

[39] Huang W. Dendrimer-encapsulated metal nanoparticles: Synthesis and application in catalysis. In: Current Trends of Surface Science and Catalysis. Springer; 2014. pp. 65-87

[40] Myers VS, Weir MG, Carino EV, Yancey DF, Pande S, Crooks RM. Dendrimer-encapsulated nanoaprticles: New synthetic and characterization methods and catalytic applications. Chemical Science. 2011;**2**:1632-1642

[41] Karakhanov E, Maximov A, ZolotukhinaA.Heterogeneousdendrimer-based catalysts. Polymers. 2022;**14**:981

[42] Scott RWJ, Wilson OM, Crooks RM. Synthesis, characterization, and applications of dendrimer-encapsulated nanoparticles. The Journal of Physical Chemistry. B. 2005;**109**:692-704

[43] Niu Y, Crooks RM. Dendrimer-encapsulated nanoparticles and their applications to catalysis. Comptes Rendus Chimie. 2003;**6**:1049-1059

[44] Nemanashi M, Noh J-H, Meijbbom R. Dendrimers as alternative templates and pore-directing agents for the synthesis of micro- and mesoporous materials. Journal of Materials Science. 2018;**53**:12663-12678

[45] Buhleier EW, Wehner W, Vögtle F. Cascade -and nonskid-chain-like. Syntheses of Molecular Cavity Topologies, Synthesis. 1978;**2**:155-158

[46] Tomalia DA, Baker H, Dewald JR, Halls M, Kallos G, Martin S, et al. A new class of polymers: Starburst-dendritic macromolecules. Polymer Journal. 1985;**17**:117-132

[47] Newkome GR, Yao Z-Q, Baker GR, Gupta K. Micelles. Part 1. Cascade molecules: A new approach to Micelles. A [27]-arborols. The Journal of Organic Chemistry. 1985;**50**:2004-2004

[48] Hawker CJ, Fréchet JMJ. Preparation of polymers with controlled molecular architecture: A new convergent approach to dendritic macromolecules. Journal of the American Chemical Society. 1990;**112**:7638-7647

[49] Sandoval-Yañez C, Rodriguez C. Dendrimers: Amazing platforms for bioactive molecule delivery systems. Materials. 2020;**13**:570

[50] Balzani V, Ceroni P, Gestermann S, Kauffmann C, Gorka M, Vögtle F. Dendrimer as fluorescent sensors with signal amplification. Chemical Communications. 2000:853-854

[51] Astruc D, Boisselier E, Ornelas C. Dendrimers designed for functions: From physical, photophysical, and supramolecular properties to applications in sensing, catalysis, molecular electronics, photononics, and nanomedicine. Chemical Reviews. 2010;**110**:1857-1959

[52] Wazir MB, Daud M, Ali F, Al-Harthi MA. Dendrimer assisted dye-removal: A critical review of adsorption and catalytic degration for wastewater treatment. Journal of Molecular Liquids. 2020;**315**:113775

[53] Astruc D, Chardac F. Dendritic catalysts and dendrimers in catalysis. Chemical Reviews. 2001;**101**:2991-3024

[54] Thomas JM. The enduring relevance and academic fascination of catalysis. Nature Catalysis. 2018;**1**:2-5

[55] Gao C, Lyu F, Yin Y. Encapsulation of metal nanoaprticles for catalysis. Chemical Reviews. 2020;**121**:834-881

[56] Niu Y, Crooks RM. Preparation of dendrimer-encapsulated nanoparticles using organic solvents. Chemistry of Materials. 2003;**15**:3463-3467

[57] Rangasamy R, Lakshmi K, Selvaraj M. Synthesis of ultrafine AuPd bimetallic nanoaparticls using a magnetic-cored poly(propyleneimine) dendrimer template and its sustainable catalysis of the Suzuki coupling reaction. New Journal of Chemistry. 2021;**45**:14227-14235

[58] Crooks RM, Zhao M, Sun L, Chechik V, Yeung LK. Dendrimer-encapsulated metal nanoparticles: Synthesis, characterization, and application to catalysis. Journal of the American Chemical Society. 2001;**34**:181-190

[59] Zhao M, Crooks RM. Preparation of Cu nanoclusters within dendrimer templates. Journal of the American Chemical Society. 1998;**120**:4877-4878

[60] Wang H, Rempel GL. Bimetallic dendrimer-encapsulated nanoparticle catalysts. Polymer Reviews. 2016;**56**:486-511

[61] Chechik V, Zhao M, Crooks RM. Self-assembled inverted micelles prepared from a dendrimer template: Phase transfer of encapsulated guests. Journal of the American Chemical Society. 1999;**121**:4910-4911

[62] Zhao M, Crooks RM. Homogeneous hydrogenation catalysis using monodispersed, dendrimer-encapsulated Pd and Pt nanoparticles. Angewandte Chemie, International Edition. 1999;**38**:364-366

[63] Yeung LK, Crooks RM. Heck heterocoupling within a dendritic nanoreactor. Nano Letters. 2001;**1**:14-17

[64] Zhao M, Tokuhisa H, Crooks RM. Interactions between organized, surface-confined monolayers and liquid-phase probe molecules. 5. Molecule-sized gates based on surface-confined dendrimers.

Angewandte Chemie (International Ed. in English). 1997;**36**:2596-2598

[65] Lang H, May RA, Iverson BL, Chandler BD. Denrimer-encapsulated Nanoaprticles precursors to supported platinum catalysts. Journal of the American Chemical Society. 2003;**125**:14832-14836

[66] Nemanashi-Maumela M, Nongwe I, Motene RC, Davids BL, Meijboom R. Au and Ag nanoparticles encapsulated within silica nanospheres using dendrimers as dual templating agent and their catalytic activity. Molecular Catalysis. 2017;**438**:184-196

[67] Yamamoto K, Imaoka T, Tanabe M, Kambe T. New horizon of nanoparticle and cluster catalysis with dendrimers. Chemical Reviews. 2020;**120**:1397-1437

[68] Seretis A, Diamantopoulou P, Thanou I, Tzevelekidis P, Fakas C, Lilas P, et al. Recent advances in ruthenium-catalyzed hydrogenation reactions of renewable biomass-derived Levulinic acid in aqueous media. Frontiers in Chemistry. 2020;**8**:1-22

[69] Montgomery TP, Johns AM, Grubbs RH. Recent advancements in stereoselective olefin metathesis using ruthenium catalysts. Catalysts. 2017;**7**:87

[70] Saadatjou N, Jafari A, Sahebdelfar S. Ruthenium nanocatalysts for ammonia synthesis: A review. Chemical Engineering Communications. 2015;**202**:420-448

[71] Maumela M, Sanette S, Meijboom R. Heterogeneous Ru catalysts as the emerging potential superior catalysts in the selective hydrogenation of bio-derived Levulinic acid to γ-valerolactone: Effect of particle size, solvent, and support on activity, stability, and selectivity. Catalysts. 2021;**11**:292

[72] Lafaye G, Williams CT, Amiridis MD. Synthesis and microscopic characterization of dendrimer-derived Ru/Al_2O_3 catalysts. Catalysis Letters. 2004;**96**:43-47

[73] Lafaye G, Siani A, Marécot A, Amiridis MD, Williams CT. Particle size control in dendrimer-derived supported ruthenium catalysts. The Journal of Physical Chemistry. B. 2006;**110**:7725-7731

[74] Azzam MO, Al-Tarazi M, Tahboub Y. Anodic destruction of 4-chlorophenol solution. Journal of Hazardous Materials. 2000;**75**:99-113

[75] Jiang S, Ni H, Li P, Wang J, Ren H. Metal/N-doped carbon (metal = Ag, Cu, Ni) nanocatalysts for selective hydrogenation of 4-nitrophenol. Catalysis Communications. 2021;**152**:106280

[76] Megharaj M, Pearson HW, Venkateswarlu K. Toxicity of phenol and three nitrophenols towards growth and metabolic activities of *Nostoclinckia* isolated from soil. Archives of Environmental Contamination and Toxicology. 1991;**21**:578-584

[77] Vaidya MJ, Kulkarni SM, Chaudhari RV. Synthesis of p-aminphenol by catalytic hydrogenation of p-nitrophenol. Organic Process Research and Development. 2003;**7**:202-208

[78] Antonels NC, Meijboom R. Preparation of well-defined dendrimer encapsulated ruthenium nanoparticles and their evaluation in the reduction of 4-nitrophenol according to the Langmuir-Hinshelwood approach. Langmuir. 2013;**29**:13433-13442

[79] Murugan E, Pakrudheen I. Efficient amphiphilic poly(propylene imine)

dendrimer encapsulated ruthenium nanoparticles for sensing and catalysis applications. Science of Advanced Materials. 2014;**6**:1-11

[80] Marvin KA, Thadani NN, Atkinson CA, Keller EL, Stevenson KJ. Preparation and catalytic evaluation of ruthenium-nickel dendrimer encapsulated nanoparticles *via* Intradendrimer redox displacement of nickel nanoparticles. Chemical Communications. 2012;**48**:6289-6291

[81] Wang Y, Peng X. RuRh bimetallic nanoparticles stabilized by 15-membered macrocycles-terminated poly(propylene imine) dendrimer: Preparation and catalytic hydrogenation of nitrile—Butadiene rubber. Nano-Micro Letters. 2014;**6**:55-62

[82] Nemanashi M, Noh J-H, Meijboom RM. Hydrogenation of biomass-derived Levulinic acid to Valerolactone catalysed by mesoporous supported dendrimer-derived Ru and Pt catalysts: An alternative method for the production of renewable fuels. Applied Catalysis. 2018;**550**:77-89

[83] Antonels NC, Meijboom RM. Preparation of well-defined dendrimer encapsulated ruthenium nanoparticles and their application as catalyst and enhancement of activity when utilised as SCILL catalysts in the hydrogenation of citral. Catalysis Communications. 2014;**57**:148-152

[84] Antonels NC, Williams MB, Meijboom RM. Well-defined dendrimer encapsulated ruthenium SCILL catalysts for partial hydrogenation of toluene in liquid-phase. Journal of Molecular Catalysis A: Chemical. 2016;**421**:156-160

[85] Peng X, Pan Q, Lu X. Regioselective catalyzed modification of poly(methylhydro)siloxane using RuRh and RuPt bimetallic dendrimer-encapsulated nanoparticles. Journal of Applied Polymer Science. 2011;**122**:334-341

[86] Gu Y, Wu G, Hu XF, Chen DA, Hansen T, Loye H-C, et al. PAMAM-stabilized Pt-Ru nanoparticles for methanol electro-oxidation. Journal of Power Sources. 2010;**195**:425-434

[87] Maiyalagan T. Pt-Ru nanoparticles supported PAMAM dendrimer functionalized carbon nanofiber composite catalysts and their application to methanol oxidation. Journal of Solid State Electrochemistry. 2009;**13**:1561-1566

[88] Karakhanov E, Maximov A, Zolotukhina A, Kardasheva Y, Talanova M. Thermo-responsive ruthenium dendrimer-based catalysts for hydrogenation of aromatic compounds and phenols. Journal of Inorganic and Organometallic Polymers. 2016;**26**:1264-1279

[89] Kashani SH, Moghadam M, Tangestaninejad S, Mirkhani V, Mohammadpoor-Baltork I. Ruthenium nanoparticles immolized on nano-silica functionalized with thiols-based dendrimer: A nanocomposite material for oxidation of alcohols and epoxidation of alkenes. Catalysis Letters. 2018;**148**:1110-1123

[90] Liu D, Jesús YML-D, Monnier JR, Williams CT. Preparation, characterization, and kinetic evaluation of dendrimer-derived bimetallic Pt-Ru/SiO_2 catalysts. Journal of Catalysis. 2010;**269**:376-387

[91] Keshtiara P, Hadadzadeh H, Daryanavard M, Mousavi N, Dinari M. New dendrimers containing ruthenium nanoparticles as catalysts for hydrogenation of citral to

3,7-dimethyloctanol. Materials Chemistry and Physics. 2020;**249**:122962

[92] Karakhanov E, Maximov A, Zolotukhina A, Mmadli A, Vutolkina A, Ivanov A. Denrimer-stabilized Ru nanoparticles immobilized in organo-silica for hydrogenation of phenols. Catalysts. 2017;7:86

[93] Maximov AL, Zolotukhina AV, Mamedli AA, Kulikov LA, Karakhanov EA. Selective levulinic acid hygrogenation in the presence of hybrid dendrimer-based catalysts. Part I. Monometallic. ChemCatChem. 2018;**10**:222-233

[94] Calderón JC, Calvillo L, Lázaro MJ, Rodríquez JL, Pastor E. Effect of dendrimer generatioin used in the synthesis of Pt-Ru nanoaprticles supported on carbon nanofibers on the catalytic activity towards methanol oxidation. Energies. 2017;**10**:159

[95] Karakhanov EA, Maximov AL, Zolotukhina AV, Terenina MV, Vutolkina AV. Nanoheterogeneous ruthenium-containing catalysts based on dendrimers in the hydrogenation of aromatic compounds under two-phase condition. Petroleum Chemistry. 2016;**56**:491-502

Section 2

Device Characterizations

Chapter 3

Inertness and Other Properties of Thin Ruthenium Electrodes in ReRAM Applications

Amrita Chakraborty, Mohammad Al-Mamun and Marius Orlowski

Abstract

Building nonvolatile memory such as resistive random access memory (ReRAM) directly into a CMOS backend (BEOL) would reduce latency in connectivity-constrained devices and reduce chip's footprint by stacking non-volatile memory (NVM) on top of the logic circuits. This co-integration is facilitated by a broad commonality between ReRAM and BEOL as both rely on the same basic metal–insulator–metal (MIM) structure. One good candidate for a ReRAM cell is the Cu/TaO_x/Pt device. As platinum (Pt) is not an economic choice, a BEOL-compatible replacement is desirable. A good candidate to replace Pt electrode is ruthenium (Ru), currently being used as a liner/diffusion barrier in sub-15 nm technology nodes and soon to supplant tungsten as via, and copper (Cu) as interconnect materials. We report on extensive characterization of a Cu/TaO_x/Ru device and compare its performance and reliability with extant ReRAM devices. Against the background of well-characterized non-Ru ReRAM devices, Cu/TaO_x/Ru cell constitutes a micro-laboratory for testing a wide range of Ru properties with the Cu nanofilament as a probe. Since the temperature of the cell can be controlled internally from 27°C to ~1100°C, thin Ru layers can be subjected to much more comprehensive tests than it is possible in the interconnect MIM structures and reveal and confirm interesting material properties, including the impact of embedment.

Keywords: chemical inertness, surface roughness, thermal conductivity, grain nucleation, columnar grains, stopping power, diffusion barrier, melting temperature, layer embedment, surface nucleation, adatom surface diffusion

1. Introduction

This chapter looks at the material properties of thin Ru films from the perspective of their suitability for use as an inert electrode in ReRAM cells, such as Cu/TaO_x/Ru [1], that lends itself to be readily integrated in the CMOS back-end-of-line (BEOL) [2]. Advanced interconnects at 10 nm half-pitch, in order to overcome the scaling issues with Cu interconnects, increasingly pin hopes on ruthenium metallization, not only as a liner and diffusion barrier for Cu, but also as a stand-alone material for contact plug, via, and even interconnect lines [3–5]. Such integration of ReRAM memory into BEOL has the potential of reducing the latency in connectivity-constrained computational devices and of bringing logic and memory closer together [6–8]. This co-integration is rendered possible and viable as low-k dielectrics and Cu/Ru/Co interconnect lines already prefigure a ReRAM device as a MIM structure, and the crossbar architecture of a typical two-terminal ReRAM memory array consists of electrode lines that may serve as interconnect lines and vice versa with metal vias being replaced by ReRAM switching layer. The replacement of the via material with the switching layer leads to the concept of formation of conductive vias on the fly, by an application of an electric pulse after the IC chip has been already manufactured [9]. Building NVM directly into a CMOS low-k interconnect back-end, would not only reduce latency in connectivity constrained computational devices but also reduce the footprint of a chip by stacking memory on top of the logic. Upon successful integration, the interconnect information bottleneck could be untied and morphed into several system architectures using the same basic universal hardware platform.

Given the auspicious circumstances, we elucidate in this chapter the material properties of Ru in the light of comparison of the performance of a Cu/TaO_x/Ru device with its well-researched ReRAM counterparts. In addition, as the ReRAM devices, such as Cu/TaO_x/Ru, can be intrinsically and locally heated up to ca 1100°C, thin Ru layers can be subjected to much more comprehensive tests than it is possible in the interconnect MIM structures to reveal material properties of as a stand-alone and as an embedded Ru material.

The ReRAM devices, by themselves, have recently been of great interest to both industry and academia as a potential replacement for a stand-alone volatile dynamic random-access memory (DRAM) and nonvolatile flash memories (NVM) that are nearing the end of their dimensional scaling roadmaps [7]. These two-terminal devices exhibit figure eight-like pinched current–voltage (I–V) hysteresis switching between a high resistance OFF state (R_{off}) and a low resistance ON-state (R_{on}) with memristive characteristics [10–12]. Resistive memory cells also hold promise for neuromorphic applications [13–15]. A specific subcategory of a ReRAM device, the conductive bridge random access memory (CBRAM) is being extensively explored as a promising candidate for a resistive memory device [16] and serves here as a probe into Ru property materials. In general, a CBRAM device consists of an active anode, an insulating layer, and an inert cathode. The anode consists of Cu, Ag, or Ni, which can dissolve in the solid electrolyte layer [17, 18]. The resistive switching (RS) behavior can be explained by a formation and rupture of a Cu conductive filament (CF). The insulating layer, a dielectric, also called the switching layer, such as Ta_2O_5, TaO_x, GeSe, GeS_2, Al_2O_3, and SiO_2 [19], allows for ion and defect electromigration when an electric field is applied. When a positive voltage is applied to the Cu electrode, Cu^+ cations are generated according to the redox reaction, $Cu \leftrightarrow Cu^+ + e^-$, which dissolve, and migrate into the solid electrolyte [20]. Cu^+ cations, after traversing the switching layer under the influence of an electric field, are electrochemically reduced on the Pt cathode which acts as an effective diffusion barrier for Cu atoms. As more Cu atoms accumulate, a nanoscale conductive filament (CF) is formed, shorting the two electrodes in so-called FORM and SET processes. The FORM operation describes the formation of the CF for the first time in a virgin memory cell. The SET operation is the restoration of an already-formed conductive filament after it had been partially ruptured in the so-called RESET operation. The ON state is characterized by the resistance of the Cu CF, R_{on}, and when the CF is ruptured the device reverts to a high resistance OFF-state, characterized by the off-resistance, R_{off}. The ON/OFF ratio of CBRAM is usually significantly higher than 10^3 [21] and potentially allows multilevel switching in a single memory cell to store more than one bit of data [22]. To date, Cu/TaO_x/Pt based devices have proven to be one of the more popular types of CBRAM devices due to numerous reports of excellent unipolar and bipolar switching characteristics, device performance, retention, reliability, endurance, and yield [23, 24]. Commercialization of non-volatile memory products based on RS devices derived from a Cu/TaO_x/Pt cell has also been reported [25].

Against this backdrop, the main focus of the chapter lies on the Cu/TaO_x/Ru device. As already mentioned, Ru is a good candidate to replace Pt and has been already deployed in the earlier CMOS BEOL technology nodes supplanting Ta or TaN as the liner material [26, 27]. Ru is ca. 45 times less expensive than Pt but has similar inertness properties as Pt. Pt and Ru are both transition metals with almost identical outer shell structure: The electron configuration for Ru is [Kr]$4d^7 5s^1$ for Pt [Xe] $4f^{14}5d^9 6s^1$, where Kr and Xe denote the noble gases, Krypton and Xenon, respectively. Ru has a single electron in the fifth orbital and 15 electrons in the fourth orbital, while the larger Pt atom has a single electron in the sixth orbital and 17 electrons in the incomplete fifth orbital. In addition, the Ru-Cu phase diagram shows negligible solid solubility between the two elements, even at 900°C, rendering Ru an excellent diffusion barrier for Cu diffusion [28]. The work function of Ru of 4.75 eV is comparable to that of Cu 4.7 eV and much lower than that of Pt of 6.35 eV which has the highest work function among metals. The material properties of metals under consideration in this investigation are summarized in **Table 1**. The work function difference between Cu and Pt/Ru is partly responsible for the higher forming and set voltages in Ru devices as compared with Pt devices.

The chapter is organized as follows: In Section 2, we describe briefly the fabrication process of the ReRAM cells employed here. In Section 3 we establish the characteristic features of the electrical performance of a Cu/TaO_x/Pt device, serving as our base line device, for comparisons with its counterparts, the Ru and cobalt (Co) devices. In Section 4 we discuss the analogous electric characteristics of the Cu/TaO_x/Ru, or Ru device for short. In this framework, we discuss how the performance of the Ru device differs from the performance of derivative devices and how these differences in performance can be related to material properties of Ru. In Section 5, we analyze how the surface roughness (SR) properties of the metals, established by Atomic Force Microscope (AFM) measurements, have an impact on the electric performance of the respective devices. In Section 6, we present results on reliability and endurance tests of the three devices, which show that Ru shows markedly degraded reliability behavior compared with the Pt devices. In Section 7, we present evidence that different embedment scenarios of the same identical Ru device, lead to significantly different electrical performance and reliability of Ru devices. Finally, in Section 8, we present x-ray diffraction (XRD) studies of the pertinent layer systems annealed at elevated

	Work Fct. [eV]	Therm. Cond. [W/mK]	Melt. Temp. [°C]	S.R. at 27°C [nm]
Cu	4.7	386	1084	2.7
Pt	6.3	69	1770	1.8
Ru	4.7	116	2334	1.5
Co	5.0	69	1495	2.3

Table 1.
Material properties of metals: Work function, thermal conductivity, and melting temperature, and surface roughness at 27°C after EBPVD deposition of metals used in our resistive switching devices.

temperatures and show that at annealing temperatures higher than 425°C, interlayer reactions in terms of formation of ruthenium and copper silicide occur, and lead to degraded functionality, particularly of the Ru devices.

2. Fabrication of ReRAM devices

Figure 1a shows the optical micrograph of the crossbar architecture and the device cross section of a typical Cu/TaO_x/Pt resistive switch. The devices are fabricated in a crossbar array on a thermally oxidized Si substrate with a SiO_2 thickness of 730 nm. Both metal electrodes and the metal oxide were deposited by electron beam physical vapor deposition (EBPVD) and patterned by layer lift-off technique. A thin titanium (Ti) layer of 25 nm was used between Pt/Ru/Co and SiO_2 to improve the adhesion of the Pt/Ru/Co layer. The thickness of the TaO_x switching layer is 25 nm. The width of metal lines varies between 5 and 35 μm resulting in rectangular device areas of the device in the range of (5 to 35) x (5 to 35) μm^2. The distance between the metal lines is 150 μm. All metal layers (Cu, Pt, Ru, Co, Cr, Ti) were deposited by EBPVD in a Kurt Lesker e-beam PVD-250 chamber. I–V characteristics were measured by a Keithley 4200-SCS at room temperature. The oxygen-deficient TaO_x ($x \approx 1.9$) was deposited also in the PVD-250 chamber by evaporating Ta_2O_5 pellets without O_2 injection into the evaporation chamber. The thicknesses of Ru, Pt, Co, and Cu layers are 50, 50, 50, and 150 nm, respectively. The details of the sample manufacturing process have been given in [21, 29] where the role of the stoichiometry parameter x in TaO_x on resistive switching properties is being discussed in detail.

Figure 2 shows the cross-section view of five devices that have been manufactured for this study. **Figure 2a** shows the base line device of Cu/TaO_x/Pt including the two derivative devices, where Pt has been replaced by Ru and Co electrodes. **Figure 2b** shows a different embedment of an identical Cu/TaO_x/Ru device as in **Figure 2a**, where an additional TaO_x layer has been inserted below the device proper, between the 730 nm thick field oxide and the Ti glue layer. **Figure 2c** shows still another embedment case of the Ru device, where the Ti glue layer has been replaced by a Chromium (Cr) glue layer. The additional 30 nm thick TaO_x layer, and the Cr layer have also been deposited by EBPVD.

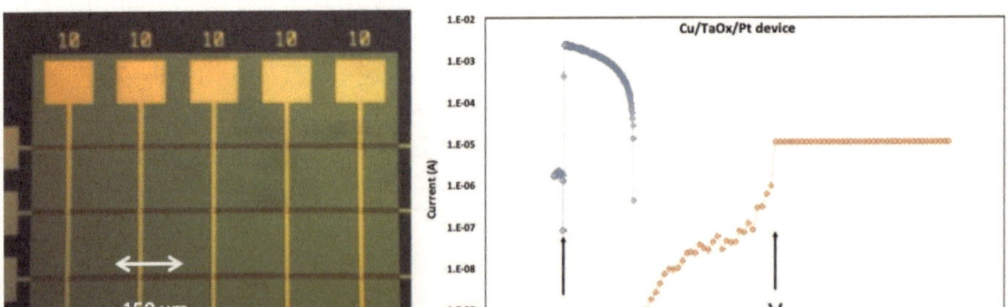

Figure 1.
*(a) Optical micrograph of the crossbar architecture of the Cu/TaO_x/Pt, Cu/TaO_x/Ru, and Cu/TaO_x/Co devices. (b) Set and reset operation of the Cu/TaOx/Pt device with I_{cc} = 10 μA. The I-V characteristics of Cu/TaO_x/Ru and Cu/TaO_x/Co devices look very similar. The critical switching voltages, V_{form}, V_{set}, and V_{res} for the Cu/TaO_x/Pt, Cu/TaO_x/Ru, and Cu/TaO_x/Co devices are summarized in **Table** 2.*

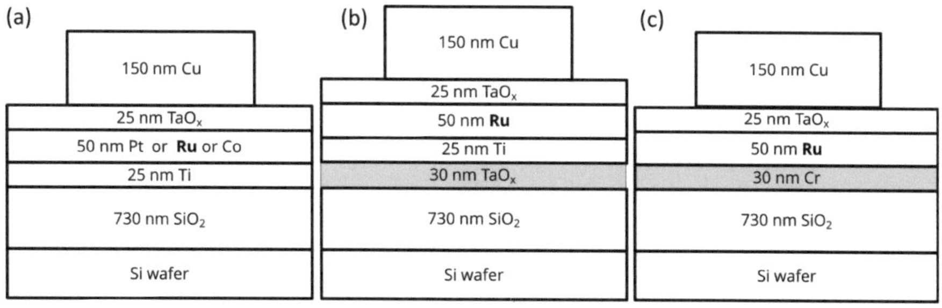

Figure 2.
Cross-sections of the five ReRAM devices. (a) Baseline Cu/TaOx/Pt device with two derivative devices, where Pt is replaced by Ru and Co. (b) Identical Ru device as in (a) but with additional TaO_x layer inserted between Ti glue layer and 730 nm thick field oxide. (c) another case of embedment of the Ru devices where the Ti glue layer has been replaced by a Cr glue layer.

3. Electrical properties of ReRAM cell (Cu/TaO_x/Pt)

In this section, we establish the basic device properties of the Cu/TaO_x/Pt device. Those properties will serve as a base line comparison for the performance of the Cu/TaO_x/Ru device. The analysis of the differences between those devices and the Cu/TaO_x/Co device allow us to assess the material properties of Ru. When a positive voltage applied to the Cu electrode of a Cu/TaO_x/Pt is swept at a constant voltage ramp rate rr [V/s], the current will remain substantially zero until a critical voltage, V_{set} is reached, at which a Cu CF is formed connecting the Cu and Pt electrodes with each other, and the cell switches from a high resistive state (HRS) characterized by R_{off} (1–900 MΩ) to a low resistive state (LRS) characterized by R_{on} (70Ω–15 kΩ), yielding a ratio of $R_{off}/R_{on} \approx 10^3$–$10^7$. When the voltage is swept back, ohmic behavior is observed until a negative voltage V_{res} is reached when the CF is ruptured and the current collapses to a very small value. A typical log-scale I-V switching characteristic of the Pt device is shown in **Figure 1b**. When the cell is set to an ON-state for the very first time, one speaks of a forming operation, characterized by the forming voltage, V_{form}. Since, during the set operation relatively few Cu atoms have to be added to pluck the gap of the ruptured filament, V_{set} is usually substantially smaller than V_{form}. For Cu/TaO_x/Pt devices, we find distribution of V_{form} with a mean, $V_{form,m}$ = 4.5 V and a standard deviation of σ = 0.6 V, the V_{set} distribution with a mean, $V_{set,m}$ = 2.8 V, and standard deviation, σ =0.6 V. The rupture of the CF is triggered mainly by Joule's heating at a critical current $I_{res} = V_{res}/R_{on}$. The V_{res} distribution of our Cu/TaO_x/Pt devices is characterized by $V_{res,m}$ = −0.9 V and σ =0.3 V. The electric characteristics in terms of threshold voltages for all devices are summarized in **Table 2**. These distributions apply roughly both to a multitude of devices as well as to a single device that has

	$V_{form,m}$ [V]	σ_{form} [V]	$V_{set,m}$ [V]	σ_{set} [V]	$V_{res,m}$ [V]	σ_{res} [V]
Cu/TaO_x/Pt	4.5	0.6	2.8	0.6	−0.9	0.3
Cu/TaO_x/Ru	7.3	0.7	4.4	0.8	−3.4	0.7
Cu/TaO_x/Co	4.7	2.2	2.7	0.9	−1.0	0.6

Table 2.
Critical switching voltages, V_{form}, V_{set}, and V_{res} for the Cu/TaO_x/Pt, Cu/TaO_x/Ru, and Cu/TaO_x/Co devices along with their sigma distribution values.

been switched repeatedly. The endurance of the Pt devices depend on the I_{cc} current during the set operation. For not too low I_{cc} (>10 μA) and not too high I_{cc} (<5 mA), the device can be switched very often. Typically, in a few cases, a device was switched for 220 times, and it could still be switched even further. In most instances, a compliance current I_{cc} is imposed, lest the devices be damaged. No I_{cc} current limitation is applied during the reset operation. R_{on} of the LRS state depends on I_{cc} via Eq. (1), where the exponent n for cation filaments is very close to unity.

$$R_{on} = A/I_{cc}{}^{n} \tag{1}$$

The R_{on}–I_{cc} relation in Eq. (1) has been reported to be valid for numerous anode/electrolyte/cathode material systems [30–32] with n ≈ 1. For Pt devices we obtain A = 0.5 V and n = 1.01. In ref. [1] it has been shown that the constant A in Eq. (1) is universally correlated to the minimum set voltage for all metallic conductive filaments reported so far. During the reset operation, the ReRAM cell is exposed to large currents which raises the temperature of the filament by depositing Joule's heat, primarily, in the Cu CF. When critical temperature in the filament has been reached, the Cu atoms begin to diffuse out and the filament is ruptured. The reset currents defined as $I_{res} = V_{res}/R_{on}$ are typically in a few mA range. Hence, most of the Joule's heat, Q_{JH} deposited during the reset operation is given by Eq. (2)

$$Q_{JH} = \int_0^{t_{res}} \frac{V^2(t)}{R_{on}} dt = \int_0^{V_{res}/rr} \frac{rr^2 \times t^2}{R_{on}} dt = \frac{V_{res}^3 \times I_{cc}}{3 \times rr \times K} \tag{2}$$

where V_{res} is the reset voltage, $t_{res} = V_{res}/rr$ is the reset time. The reset ramp rate, rr, and the compliance current, I_{cc}, may be used to control amount of heat deposited in the cell. Depending on chosen values for I_{cc} and rr, Q_{JH} can vary from 3 to 60 μJ [33]. The possibility of controlling the dissipated heat in the filament in terms of compliance current, I_{cc}, and ramp rate, rr will be important for the characterization and analysis of the electrical performance of the Ru, Pt, and Co devices.

The geometrical shape of the filament has considerable implication of the reset operation. Three different shapes of Cu CF are shown in **Figure 3**. In **Figure 3a**, a CF in the shape of a truncated cone with a sharp constriction at the top is shown. The bulk of its resistance is concentrated at the tip of the cone. Filaments with a sharp

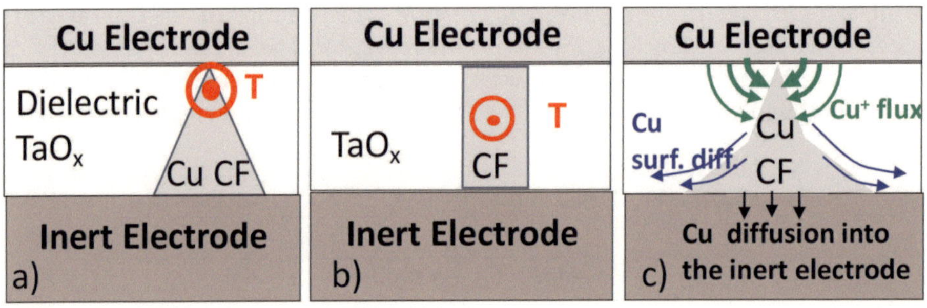

Figure 3.
Three different shapes of the Cu filament that may occur at different SET operation conditions with location of the highest temperature during RESET operation indicated. (a) Highly resistive Cu CF with sharp constriction at the top of the cone. (b) a low resistance Cu CF of cylindrical shape. (c) Constructive and destructive Cu fluxes during formation of Cu CF.

constriction at the top of the cone easily can be ruptured since the maximum Joule's heat is deposited at the tip and leads there to a high local temperature, which, in turn, triggers Cu out diffusion and formation of a gap in the filament. When the filament is ruptured, the HRS state of the cell is being restored. In contrast, the cylinder shape of a robust Cu filament shown in **Figure 3b** leads to much lower resistance and the temperature hot spot is moving away from the Cu/TaO_x interface to midway between the electrode interfaces, where it is much more difficult to reach high temperature and cause out-diffusion of a larger number of Cu atoms to form a gap. Experimentally, we consistently find that the rupture of a filament is more difficult at lower R_{on} resistance, that is, formed at higher I_{cc} than at high R_{on} resistance formed at low I_{cc}.

During the final stage of the set operation there are two Cu fluxes that are responsible for the formation of the filament as shown in **Figure 3c**. The constructive flux occurs at the tip of the filament where the voltage drop is the largest and creates high electric field between the filament tip and the Cu electrode, which, in turn, allows for an enhanced transport of Cu^+ ions from the Cu electrode to the filament even after establishing the initial connection between the Cu filament and the Cu electrode. The role of I_{cc} is to limit the resulting voltage drop and thus to set a limit on the electric field driving the Cu^+ ion transport. When sufficient number of Cu atoms is deposited at the tip, the resistance of the filament decreases, reducing the electric field at the tip and bringing the formation of the filament to a halt at a given I_{cc}. When I_{cc} is increased, then the electric field at the tip increases proportionally, triggering additional arrival of Cu^+ ions until the resistance of the filament drops sufficiently to reduce the electric field and, thus, to halt further Cu^+ ion transport. The other Cu flux is the Cu atom diffusion flux that weakens the base of the filament leading to a gradual change of the filament's shape from a truncated cone to cylinder-like shape. The second Cu flux has two components: (i) Lateral Cu diffusion along the TaO_x/inert electrode interface, and (ii) Cu diffusion into the inert electrode. The Cu atom diffusion flux out of the filament at the base is impacted by the thermal conductivity of the inert electrode. Ceteris paribus, the inert electrode with high heat conductivity will be able to remove heat at a higher rate than an electrode with low heat conductivity. The larger heat removal rate will result in lower attainable maximum temperature of the Cu CF. Hence, a cell with inert electrode of low thermal conductivity is bound to display enhanced Cu diffusion and structural weakening of the filament at a lower V_{res} voltages than high thermal conductivity electrode. The high temperature in the filament during the reset will trigger Cu diffusion near the base of the filament partly into the dielectric, partly along the TaO_x/inert electrode interface, and partly into the inert electrode if the inertness of the electrode be somehow compromised. The net result of those diffusion components is a weakening of the base of the cone and a gradual transformation of the shape of the filament from a sharply cone-shaped into a more cylinder-shape filament as shown in **Figure 3b**. Once the shape of the filament is sufficiently close to that of a cylinder, it becomes very resistant to rupturing, and the number of switching cycles comes to a halt.

4. Electric characteristics of the ruthenium (Cu/TaO_x/Ru)

The electrical characterization of the Ru device reveals that the device behaves very much like Pt and Co devices, albeit with some notable differences that can be all satisfactorily explained in terms of the material properties of the three metals given in **Table 1**.

Table 2 summarizes the critical switching voltages for the three devices and shows that all the critical voltages of the Ru devices are considerably higher than those of Pt devices: ΔV_{form}(Ru-Pt) = 2.7 V, ΔV_{set}(Ru-Pt) = 1.6 V, and ΔV_{res} (Ru-Pt) = 2.5 V. One notices that the work function difference $\Delta\varphi$(Ru,Cu)/q = 1.6 V between Pt and Ru corresponds exactly to ΔV_{set} = 1.6 V between Pt and Ru devices indicating that V_{set} values are being reached at the same strength of the electric field. Since the work function contributes to the internal electric field across the dielectric, this difference was to be expected. Comparing now V_{form} of Pt and Ru devices one notes that V_{form} for Ru is higher by 2.8 V than V_{form} for Pt, significantly higher than 1.6 V. This differential can be accounted in part by 1.6 eV as a difference of the work function between Pt and Ru, as discussed before.

The remaining voltage differential of 1.2 V can be attributed to the electric field enhancement factor that is correlated to the higher SR of Pt (1.8 nm) than that of Ru (1.5 nm). Higher SR, in general, causes a higher local field across the oxide at the same voltage, or in other words, the same critical field required for the Cu CF formation is reached at lower voltage in Pt than in a Ru device. In **Table 2**, one observes also that V_{form} for Co device is only slightly higher than V_{form} for Pt device, despite a work function difference of $\Delta\varphi$(Pt-Co)/q = 1.3 V. Based on the work function argument, one would expect a higher V_{form} by about 1 V for Co devices than for Pt devices. However, one notes that the surface roughness of Co (2.4 nm) is significantly higher than that of Pt (1.8 nm). (SR will be discussed in more detail in Section 8.) The higher surface roughness of Co leads to a large enhancement factor of the local electric field due to the smaller curvature radius at the tip of the asperity. For a more detailed discussion of how the field enhancement factor correlates with SR, see ref. [34]. The differences in SR explain that in order to reach a critical field required for the formation of a filament a lower voltage is needed in case of a rougher surface of Co than for smoother interfaces of Pt or Ru. Both effects help lower the V_{form} voltage of Co devices to a voltage only slightly higher than V_{form} for a Pt device. These comparisons show that for the filament formation the local electric field, and by extension, the surface roughness, is an important factor. One also notices that the surface roughness correlates well with the standard deviation of the V_{form} distributions of the three metals. The highest standard deviation observed for Co ($\sigma_{form(Co)}$ = 2.2 V) correlates well with highest surface roughness of $SR_{(Co)}$ = 2.4 nm of the cobalt devices, compared with the respective values of $\sigma_{form(Ru)}$ = 0.6 V and $SR_{(Ru)}$ = 1.5 nm for Ru, and $\sigma_{form(Pt)}$ = 0.6 V and $SR_{(Pt)}$ = 1.8 nm. for Pt devices.

Because of the work function difference between Pt and Cu of 1.6 eV, there is a considerable built-in field that favors Cu^+ ion transport toward the Pt electrode. In fact, it has been observed [35] that even at a TaO_x thickness of 25 nm, the work function difference is sufficiently high to set the device spontaneously into a conductive state after the device has previously been reset. This means that the electric field in the ruptured gap of the filament has reached the critical field of 3×10^6 V/cm [36] to set the device spontaneously. Not a single instance of spontaneous setting of the devices has been observed for the Ru and Co devices tested at the same conditions. Thus, the spontaneous forming of a ruptured filament is enabled by the high work function difference between Pt and Cu metal electrodes. Therefore, in the absence of the built-in electric field, the electric field in Ru device has to be increased considerably by application of external bias to reach the critical field strengths in the TaO_x dielectric. It has been shown [36] that a field of $\approx 2.9 \times 10^6$ V/cm has to be reached in order to establish a Cu CF across a thin TaO_x dielectric layer, independently of the nature of the inert electrode.

Next, we discuss the large difference in V_{res}, between Ru and Pt device, ΔV_{res} (Ru-Pt) = 3.4–0.9 = 2.5 V, see **Table 2**. There is a general consensus that the rupturing of the Cu filament is attributed to the partial dissolution of the filament due to the Joule's heating. The local temperature at the thinnest part of the filament can reach temperature in excess of 1084°C [37]. 1084°C is the melting temperature of Cu (see **Table 1**) and melting of Cu electrodes has been experimentally observed after excessive heating of a ReRAM cell [38]. At such high temperature, the Cu atoms of the Cu CF diffuse out, may undergo the redox reaction in TaO_x, and are returned as Cu^+ ions by the electric drift fields back to the Cu electrode or stay immobilized in TaO_x material. The dissolution of the filament, triggered primarily by the Cu out-diffusion into the TaO_x matrix, depends on the critical temperature reached in the constriction of the CF. The overall balance of heat stored in the nanofilament is given by the resistive Joule's heating of the nanofilament and via heat removal mainly by the heat sinks of the two relatively cold mesoscopic electrodes. The largest interface of the nanofilament is formed with the inert electrode (Pt, Ru, and Co) which forms the base of a truncated cone or a cylinder of the filament with the inert electrode (**Figure 3**). In this context, one notices that Ru has thermal conductivity of 119 W/Km almost twice those of Co and Pt of 69 W/Km (**Table 1**). Thus, lower thermal conductivity of the inert electrode requires less Joule's heating than an inert electrode with high thermal conductivity as the heat accumulates in the nanofilament in a shorter time than in the case of high thermal conductivity of the inert electrode. When inspecting the V_{res} voltages of the three devices in **Table 2**, one notices that the V_{res} values for Pt (−0.9 V) and Co (−1.0 V) devices are essentially the same, which correlates again very well with the same thermal conductivities of the two metals of 69 W/Km.

Since the Cu filament in all three devices serves as our probe into the analysis of the material properties of the three inert electrode metals (Ru, Pt, Co), it is instrumental to ensure that the nature of the Cu CF probe is the same in the three devices. To this end, we have measured R_{on} dependence on I_{cc} of the three devices. We find that all devices display the same dependence given in Eq. (1) with exponent n being very close to 1, and the constants A of eq.(1), being 2.0 V, 0.5 V, and 0.5 V, for Ru, Pt, and Co devices. These values correlate very well with the V_{set} values for these devices given in **Table 2**. The high A value for Ru is consistent with the interpretation given in [1] that A reflects the lowest possible set voltage of the device.

Conductive filaments can also be fingerprinted by their temperature coefficient of resistance (TCR), α, as defined by the temperature dependence of the R_{on} resistance: $R(T) = R(T_o) \times [1 + \alpha(T-T_o)]$, which sensitively depends on the strength of the Cu filament. We have extracted the TCR from resistance measurements of R_{on} of Cu CF as a function of temperature, T. We find for a weak Cu CF of $R_{on} \approx 15$ kΩ (corresponding to $I_{cc} = 10$ μA) following TCR values for the three devices: TCR (Ru device) = 0.00236 K^{-1}, TCR(Pt-device) = 0.00235 K^{-1}, and TCR(Co device) = 0.00235 K^{-1}. It can be seen that they are almost identical. The same TCR measurements have been repeated for stronger (lower resistance) Cu CFs set at three orders of magnitude higher I_{cc}, $I_{cc} = 10$ mA. For $I_{cc} = 10$ mA we obtain much lower R_{on} value for the Cu filaments of 510 Ω, 230 Ω, and 315 Ω for the Ru, Pt, and Co devices, respectively. For the three robust CFs we obtain very similar TCR values are extracted: TCR (Ru device) = 0.0035 K^{-1}, TCR(Pt device) = 0.0036 K^{-1}, and TCR(Co device) = 0.0036 K-1, which are, again, very similar. These values are typical of strong Cu filaments; for comparison the TCR of bulk Cu is 0.0039–0.004 K^{-1}. These results ensure that in all the three devices the Cu filaments have the same properties conferring, thus, confidence that the Cu CF may serve as a reliable probe into the properties of the inert electrode materials Ru, Pt, and Co.

5. Surface roughness studies of Ru, Pt, Co, and Cu

In this section, we investigate the surface roughness of Ru, Pt, Co, and Cu thin films (50 nm) as a function of anneals at elevated temperatures for two cases: free and passivated metal surfaces. The impact of SR of metal electrodes on local field enhancement, dielectric breakdown, including ReRAM performance is well known and has recently attracted a lot of interest. A review of the impact of surface roughness on numerous devices and phenomena, including gas breakdown, dielectric breakdown in RF MEMS devices, in triboluminescence, MIM structures, capacitors, and ReRAM memory cells can be found in [34]. As early as 1928, Eyring et al. [39] found that logarithms of currents of extracted electrons from asperities display linear relation with the reciprocal of electric field strength at the tip of an asperity. The asperity has been modeled as a prolate spheroid of length c and base radius b. A field enhancement factor has been derived:

$$\beta = \left(\frac{c/r}{ln\left(2\frac{b}{r}\right) - 1}\right) \tag{3}$$

where $r = b^2/c$. For micro size asperities with c = 16 μm and b = 0.7 μm one obtains an enhancement factor of $\beta = 187$. Countless studies have confirmed since then the impact of the electric field enhancement factor on various physical systems [34].

Separately from the ReRAM devices, we have deposited by EBPVD blanket layers of the four metals (Pt, Ru, Co, and Cu), each 50 nm thick, and using Rapid Thermal Annealing (RTA) annealed them at various temperatures with free and SiO_2-passivated surfaces – mimicking, thus, conditions that are likely to occur in the actual ReRAM cell during the switching events. Using Atomic Force Microscope (AFM), we have measured the surface roughness of the Pt, Ru, Co and Cu at room temperature, that is, just after their depositions by EBPVD, and after anneals using RTA of the free surfaces at 425°C for 8 min, 600°C for 5 min, and at 900°C for 2 min. As the SR of free surfaces after an anneal at 425°C, 600°C,and 900°C are not expected to be entirely representative of what may happen inside the memory cells, where the critical surfaces are interfaced with a SiO_2 or a TaO_x switching layer, we have passivated the surfaces of all metals with 30 nm SiO_2 layer and subjected the samples to the same anneals at 425°C for 8 min, at 600°C for 5 min, and at 900°C for 2 min. After the anneals are done, the 30 nm SiO_2 layer has been removed by a soft buffer oxide etchant (BOE) etch, highly selective to the metals, to expose the surface of the inert metals for the subsequent AFM measurements. The results of surface roughness measurements are summarized in **Table 3**.

Several interesting observations can be drawn from **Table 3**. In general, the surface roughness of a free surface increases with increasing anneal temperature. This increase is much larger for free than for passivated surfaces. It can be seen that an anneal at 600°C or 900°C with a passivated surface, leads to a smoothing effect of the surface compared to a free metal surface. However, occasionally, some irregularities defy the general trend. For example, the surface roughness of Co annealed at 600° is higher with passivation than without it. It is seen that at room temperature (27°C) out

	27°C	F.S. 600°C 5 min	F.S. 900°C 2 min	SiO_2 425°C 8 min	SiO_2 600°C 5 min	SiO_2 900°C 2 min
Pt	1.8	2.4	4.9	1.6	1.3	2.3
Ru	1.5	2.0	4.2	1.3	1.4	2.1
Co	2.4	2.8	14.0	7.3	5.0	2.4
Cu	2.2	4.4	12.0	4.1	3.0	6.5

Table 3.
Surface roughness of metal surfaces in nm as a function of temperature and presence or absence of a SiO_2 passivation layer. F.S. denotes a free metal surface annealed at different temperatures. SiO_2 denotes the cases in which the metal was covered by a 30 nm SiO_2 during the anneal, and the SiO_2 was removed by a soft oxide etch after the anneal [34] © The Electrochemical Society reproduced by permission of IOP Publishing. Chakraborty A, Al-Mamun M, Orlowski M. Impact of surface roughness and material properties of inert electrodes on the threshold voltages and their distributions of ReRAM memory cells" ECS J. Sol. St. Sci. Technol.2022;11:104007; DOI: 10.1149/2162-8777/ac9c91.

of the four metals the lowest surface roughness has ruthenium of 1.5 nm. The surface roughness of a free Pt surface increases moderately after the 600 and 900°C anneals to 2.4 and 4.9 nm, respectively. When, however, Ru (Pt) is covered during the anneal with SiO_2, the surface roughness after 600°C anneal is measured to be only 1.4 nm (1.3 nm) and climbs to 2.1 nm (2.3 nm) at 900°C. Thus, the surface roughness of Ru and Pt with passivation, when annealed at 600°C, are even smaller than at 27°C. The low increase of surface roughness of Ru with temperature compared to other metals is consistent with the fact that Ru has the highest melting temperature (2334°C) of the four metals, see **Table 1**. The highest surface roughness of a free surface at 27°C is found for Cu and Co to be 2.2 nm and 2.4 nm, respectively. The surface roughness of Co increases to 2.8 nm after an anneal of 600°C for a free surface, but becomes very large for free surfaces annealed at 900°C yielding 14.0 nm for Co and 12 nm for Cu. One observes that with SiO_2 passivation the surface roughness increase is much more moderately and is capped at 5.0 nm for Co and by 6.5 nm for Cu. It is noted that the highest surface roughness for Cu and Co and correlates well with their low melting temperatures, Cu (1084°C) and Co(1495°C). The SR of Pt displays an intermediate behavior between Ru and Co. The SR of Pt is smaller than that of Co and Cu but is higher than that of Ru. After anneals at 425, 600, and 900°C with passivated surfaces, the SR Pt increases moderately slightly more than that of Ru. This correlates well with the melting temperature of Pt (T_m = 1770°C) and that of Ru (T_m = 2335°C). These results for surface roughness apply to thin layers deposited by EBPVD. Similar trends are expected to hold for other deposition techniques. It is known, for example, that SR of Pt films deposited by plasma enhanced atomic layer deposition [40] is lower than that of PVD and lies usually below 1 nm. Nevertheless, because the metals are deposited by the same method where the deposition rate has been kept the same for all four metals, the surface roughness results should be indicative of their differences in material properties.

Figure 4 shows the AFM pictures of free surfaces of the metals at room temperature (27°C) and after an RTA anneal at 600°C for 5 min. One can see that the Ru surface is most stable of all metals, while the surface of Cu undergoes a significant reconstruction. The cases of Co and Pt lie in-between, showing a strong correlation between surface roughness and the corresponding melting temperature. Thus, with the exception of Cu, the topology of surface and surface roughness of the metals do not change appreciably. In case of Cu, however, a complete surface reconstruction is

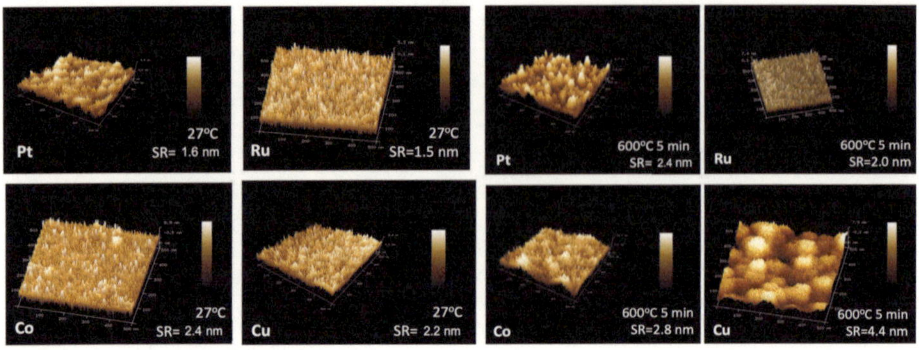

Figure 4.
AFM pictures of free surfaces of Pt, Ru, Co, and Cu at 27°C and at annealed at 600°C for 5 min with the average surface roughness indicated [34] © The Electrochemical Society reproduced by permission of IOP Publishing. Chakraborty A, Al-Mamun M, Orlowski M, "Impact of Surface Roughness and Material Properties of Inert Electrodes on The Threshold Voltages and Their Distributions of ReRAM Memory Cells" ECS J. Sol. St. Sci. Technol.2022;11:104007; DOI: 10.1149/2162-8777/ac9c91.

being observed. While at room temperature, the Cu surface resembles that of the inert electrode metals Pt, Ru, and Co, the Cu surface at 600°C shows formation of large clumps or hillocks of more than 100 nm size in diameter, while the finer dendrites on those hillocks appear to have the same needle-like structure as Cu at the 27°C. The similarities extend even to the surface morphology as both surfaces for Cu and Co undergo surface reconstruction, particularly severe in case of Cu and Co at 900°C. In **Figure 5**, the surface roughness of the metals is shown after being passivated and annealed at 600 and 900°C with a SiO_2 passivation layer on top. Before the AFM measurement the SiO_2 layer has been removed by a soft HF etch.

At 600°C, for Cu and Co, one can see hillock formation with SR higher than that at 27°C but significantly smaller than that of the free surfaces annealed at the same temperatures. It can also be seen that for Ru and Pt the SR is much smaller than for free surfaces and even smaller than their SR at room temperature. Thus, with respect to the surface roughness, the metals Pt and Ru may rightfully be called inert metals with respect to the thermal properties of their surfaces. After a 900°C anneal, the topography of the Ru, Pt, and Co surfaces does not change much, and the SR increases

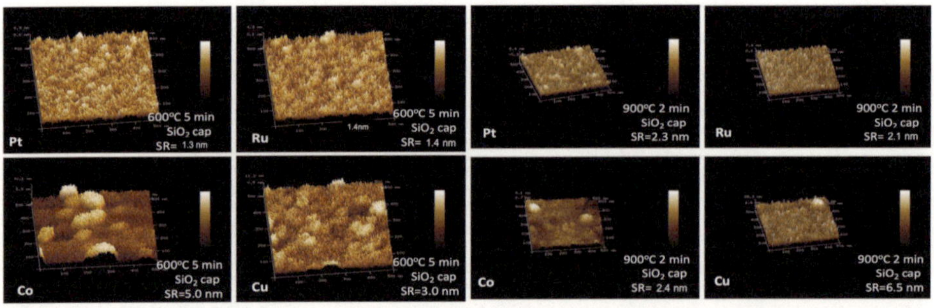

Figure 5.
AFM pictures of passivated surfaces of Pt, Ru, Co, and Cu annealed at 600 and 900°C. Before taking AFM measurements the SiO2 passivation layer was stripped by a soft etch [34] © The Electrochemical Society reproduced by permission of IOP Publishing. Chakraborty A, Al-Mamun M, Orlowski M. "Impact of Surface Roughness and Material Properties of Inert Electrodes on The Threshold Voltages and Their Distributions of ReRAM Memory Cells", ECS J. Sol. St. Sci. Technol.2022;11:104007; DOI: 10.1149/2162-8777/ac9c91.

moderately over those measured at 27°C and the anneal at 600°C. Overall, it can be concluded from **Table 3** that a passivation layer suppresses the increase of SR with increasing temperature significantly.

We have also measured the surface roughness of the switching layers SiO_2 and TaO_x, deposited in both cases by PVD directly on oxidized Si-wafers. We obtain SR at room temperature of 0.9 and 1.0 nm, for SiO_2 and TaO_x, respectively, and annealed at 600°C the SR drops to 0.53 and 0.62 nm, respectively. The reflowing property of SiO_2 and TaO_x has a significant smoothing effect on the surface roughness of the inert electrode. Our results confirm the findings that nanometer thin SiO_2 [41] and TaO_x [42] have much better reflow properties than their mesoscopic or bulk counterparts which need doping such as boron and/or phosphorous to lower the reflow temperature.

The reflow properties of SiO_2 and TaO_x layers have been confirmed by measurements of surface roughness of Ru covered with SiO_2, the electrolyte with the smallest native surface roughness. Even the mere deposition of SiO_2 has a smoothing effect. In case of Ru, the SR at room temperature is 1.5 nm, but when covered with PVD-SiO_2 the surface roughness of the SiO_2/Ru bilayer is only 1.2 nm. Since SR of SiO_2 deposited directly on the oxidized Si wafer is 0.9 nm, it can be seen that the composite surface roughness of the two layers is not additive and that SiO_2 exerts a smoothing effect. If the cumulative SR were additive, the SR would add up to 2.4 nm. Even more remarkably is the smoothing effect of SiO_2 at higher temperatures: the surface roughness of SiO_2 on Ru after the anneals of 250°C for 5 min, 600°C for 5 min, and 900°C for 2 min remains at a more or less constant value of 1.3 nm. This is contrasted with the surface roughness of a free Ru surface after the 900°C anneal when the SiO_2 layer has been removed after the anneal of 2.1 nm and the SR of a free Ru surface after 900°C anneal of 4.2 nm. Thus, dielectrics such as SiO_2 or TaO_x display even stronger surface smoothing action at elevated temperatures. These results suggest a method of reducing the surface roughness of metals by depositing a thin sacrificial layer of dielectric such as TaO_x, or SiO_2 followed by an anneal at moderate temperatures such as 425°C before removing the passivation layer by a soft BOE etch.

Finally, we wish to comment on how the surface roughness relates to the electric field enhancement factor, β, given in Eq. (3). The β enhancement factor depends largely on the curvature radius at the top of an asperity. Surface roughness, on the other hand, quantifies only the deviations of the asperities in the direction normal to the surface and does not account for the curvature radius of the tip. Nevertheless, as seen from **Figure 4**, the surface morphology of the four metals at room temperature is very similar and the respective field enhancements may still be expected to correlate well with the surface roughness as a lump parameter.

6. Degraded reliability and endurance of Cu/TaO_x/Ru device

The major drawback of the Ru devices as compared to their Pt counterparts is their limited switching capacity. While Pt devices can be switched repeatedly back and forth for at least 220 times, Ru devices are becoming non-resettable after several set-reset operations (usually less than 13), and sometimes even after the first set operation, when the set operation is performed at high I_{cc}. In some cases, it was even difficult to reset a high resistance Cu filament in a Ru device formed at I_{cc} as low as 5 μA. The failure of the Ru devices after a few switching cycles is likely to be related to the specific geometric configuration of the Cu filament which may be caused either by the diminished stopping power of the Ru electrode for Cu diffusion into the Ru film or by an increased surface diffusion as compared to that for the Pt electrode.

It is well known that even in extant ReRAM devices, including Pt devices, it is difficult and sometimes impossible to reset the cell if the cell has been set at high I_{cc}. For a low $I_{cc,}$ the shape of the CF can be approximated by a truncated cone as shown in (**Figure 3a**) where the bulk of the CF's resistance resides in the tip of the cone. Since during the reset the power dissipated in the filament is equal to $I_{res}^2 \times R_{on}$, the highest temperature is reached in the constriction at the tip of the cone, where the rupturing of the filament is easy. At high I_{cc}, the R_{on} decreases sharply by adding Cu atoms to the upper section of the cone, rendering filament's shape more and more cylindrical (**Figure 3b**). In this case, the maximum temperature, in the absence of any pronounced constriction, is reached in the middle section of the cylinder, where the low resistance filament is relatively strong. Hence, rupturing of the CF becomes difficult or impossible.

The cylindrical shape of the CF may also be obtained if the base of the cone in contact with Ru electrode erodes, see **Figure 6**. As the bottom base of the CF is reduced, while new Cu^+ ions are added from the top during the set operation, the shape of the filament will transform gradually into a cylindrical shape. There are several possible mechanisms for the erosion of the base of the filament: (i) Cu surface diffusion along the Ru interface (ii) crystallization of Ru and out-diffusion of Cu along the Ru grain boundaries, (iii) formation of Cu and Ru silicides. Those degradation mechanisms are depicted in **Figure 6a** and will be discussed further in Section 8. Whereas, with a nearly perfect stopping power, the shape of the filament in the Pt device can be assumed to remain conical with more or less sharp tip at the Cu electrode, the shape of the Cu filament in a Ru device is more cylinder-like, especially, when the Ru device was self-annealed by undergoing several set-reset switching cycles. We find also large variability of performance among Ru devices becoming noticeable after a few switching cycles. This variability may be explained by the relative location of the Cu filament with respect to the locations of the Ru grain boundaries as illustrated in **Figure 6b**. When there is no such overlap, the cell would perform well for several switching cycles. If, however, there is overlap of the base of

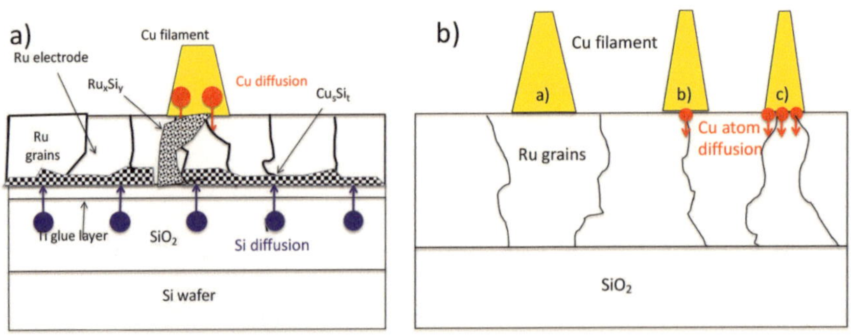

Figure 6
(a) Possible mechanisms of the hypothesized degradation of the Ru electrode. Elevated local temperatures in the immediate vicinity of the filament may cause Si diffusion into Ru, possibly along Ru grain boundaries and lead to Ru_xSi_y reactions. Cu may use the Ru grain boundaries as diffusion paths and may also undergo a silicidation reaction. (b) Relative location of the Cu filament with respect to the Ru grain boundaries likely to explain the large variability of performance see in Ru devices [35] © The Electrochemical Society reproduced by permission of IOP Publishing. Al-Mamun M, King, S, Orlowski M. "Thermal and Chemical Integrity of Ru Electrode in Cu/TaOx/ Ru ReRAM Memory Cell", ECS J. Sol. St. Technol.2019; 8: N220; DOI: 10.1149/2.0121912jss.

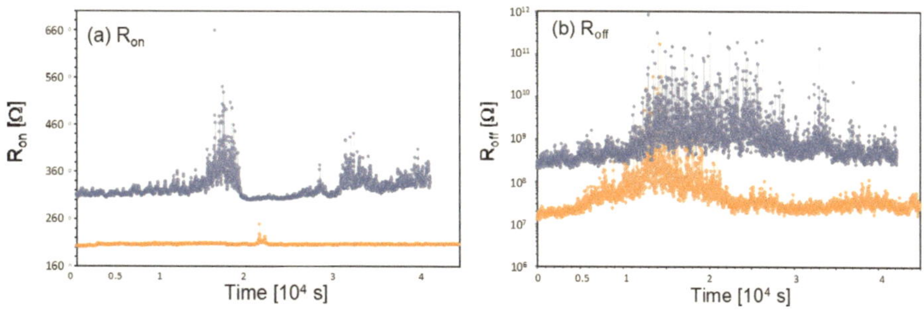

Figure 7.
The retention of Ru and Pt devices (a) for the on-state characterized by R_{on}, and (b) for the off-state characterized by R_{off} [35] © The Electrochemical Society reproduced by permission of IOP Publishing. Al-Mamun M, King, S, Orlowski M. "Thermal and Chemical Integrity of Ru Electrode in Cu/TaOx/Ru ReRAM Memory Cell", ECS J. Sol. St. Technol.2019; 8: N220; DOI: 10.1149/2.0121912jss.

the filament with one or more grain boundaries, then Cu atoms may be easily depleted from the filament via enhanced Cu diffusion along the grain boundaries.

In this context, we wish to mention that in ref. [21] other related ReRAM devices, such as Cu/TaO_x/Ti and Cu/TaO_x/Ta have been compared to the benchmark device Cu/TaO_x/Pt and were found significantly inferior to the Pt device. The Ru device is found to perform better than the Ti and Ta devices.

As already mentioned, in terms of endurance Pt devices are by far best, switching over 200 times, followed by Co devices that can be switched at optimum conditions up to 40 times, while Ru devices could be switched at best 13–14 times. The retention of both binary states characterized by R_{on} and R_{off} for all three devices is good. In **Figure** 7 the retention of the on- and off-states for Ru and Pt devices is shown.

It can be seen from **Figure** 7 that the level of noise for Ru devices is higher than for Pt devices for both bit states, characterized by R_{on} and R_{off}. While R_{off} fluctuations of Ru devices are irrelevant as the outlier lie in the tens of GΩ range, the outliers for R_{on} are more of a concern, as its variation can be as high as 75%. Such large fluctuations could pose, inter alia, a problem for the multibit storage applications and for the stability of the sense amplifiers when reading the cell. For comparison the variations of R_{on} and R_{off} in Pt devices are less than 10% as discussed in more detail in ref. [35].

Before focusing on specific mechanisms that degrade the inertness properties of Ru films, we will discuss performance changes of nominally the same devices caused by their different integration scheme, called here the "embedment" aspect.

7. Impact of embedment on the performance of Cu/TaO_x/Ru devices

To study the relative degradation of Cu/TaO_x/Ru vs. Cu/TaO_x/Pt devices, we have manufactured two nominally identical Cu/TaO_x/Ru devices (where the layers of the device proper have been processed at the same time), however embedded differently on the Si wafer as shown in **Figure 2b** and **c**. The Ru device A has the same substrate as the Pt device, that is Ti(25 nm)/SiO_2(730 nm)/Si-wafer, while the Ru device B is manufactured on the layer stack Ti(20 nm)/TaO_x(30 nm)/SiO_2(730 nm)/Si-wafer. Thus, Ru device B has an additional TaO_x-30 nm layer inserted between SiO_2 and Ti layers as shown in **Figure 2b**. We find that the Ru device B with the additional TaO_x layer shows much better performance (up to 37 switching cycles) than the Ru device

A, at all test conditions, although to a varying degree. The degree of the improvement in the switching properties between the two embedment scenarios, depends on the amount of internal Joule's heat exposure and varies with the levels of applied compliance current and ramp rate, which control the amount of heat according to Eq. (2). The highest improvement of device B over the device A is found for I_{cc} = 50 μA and rr = 0.2 V/s (high heat dissipation), and little improvement in performance between the two devices can be observed for the condition I_{cc} = 1 μA and rr = 2.0 V/s (small heat deposition) - a regime where both devices perform the best. It is found that at I_{cc} = 50 μA and rr = 0.2 V/s the Ru device A does not display resistive switching behavior at all, i.e., the device is dysfunctional. The device A cannot be reset after the filament has been formed the very first time. However, when the reset ramp rate is increased fivefold to rr = 1 V/s – thus reducing the Joule's heat dissipated in the device - the device A can be neither set or, if the set operation is eventually successful, the device cannot be reset, that is, the resistive switching cell has been permanently damaged. In contrast, device B, with the additional TaO_x, layer shows some resistive switching behavior for a few cycles (maximum 6). Hence, it can be concluded that high heating effects are detrimental to both devices. But, while Ru device B displays some, if small, degree of resistive switching behavior (5–6 switching cycles), Ru device A does not display resistive switching at all. Keeping the reset ramp rate at 0.2 V/s but now reducing the I_{cc} current from 50 μA to 1 μA, we observe improved resistive switching behavior for both devices with the frequency of the Ru devices B being twice as high (maximum 22 cycles) as for the Ru devices A (maximum 10 cycles). These results clearly demonstrate that the endurance of the device is directly related to the amount of heat deposited in the devices leading to some adverse reactions which are mitigated by the insertion of the TaO_x layer subjacent to the Ru film.

Another embedment case has been implemented when two identical devices Cu/TaO_x/Ru have been fitted with different glue layers as shown in **Figure 2c** [2]. Here, the Ti glue layer for the inert electrode has been replaced with a Cr glue layer of 30 nm. It was found that the electric performance of nominally identical devices depends sensitively on the nature of the glue layer. We find that the V_{res} value for Ru/Cr is 0.4 V larger ($V_{res(Ru/Cr)} = -3.8$ V) than for the Ru/Ti device ($V_{res(Ru/Ti)} = -3.4$ V). This could have been predicted on the grounds of four times higher thermal conductivity for Cr [94 W/m.K] than that of Ti [22 W/m·K] leading in case of Ru/Cr to much higher heat removal rate from the Cu CF during the reset operation, as discussed already in Section 4. In case of Ru/Ti, the Joule's heat generated in the filament cannot be as easily dissipated, and lingers therefore for some time around the hot spot in the filament, as in the case of the Ru/Cr device. Both embedment experiments suggest that the heat generated in the memory cell may trigger some kind of chemical or grain growth reactions that can be somewhat mitigated by the insertion of the TaO_x layer.

8. XRD studies of different embedment of Ru layers

In order to find out what kind of chemical reactions may occur in the Ru devices at various temperatures, we have manufactured among many auxiliary blanket multilayer structures on oxidized Si wafers following samples: (i) Ru(50 nm)/Si, (ii) Ru (50 nm)/SiO_2(611 nm)/Si, and (iii) Ru(50 nm)/Ti(40 nm)/TaO_x/SiO_2(611 nm)/Si that we consider here. The auxiliary samples are bound to avail some information on Ru crystallization and the presence of new materials such as silicides using the XRD technique. The XRD measurements have been performed on the three samples for three annealing conditions: (i) unannealed, that is, at 27°C, (ii) annealed at 600°C for 10 min, and (iii) annealed at 900°C for 30 min. In **Figure 8a**, a Ru crystallization peak at 42.15° can be seen on the XRD spectra for all samples with a Ru layer at 27°C and after an anneal at 600°C, and 900°C. This is in agreement with XRD studies of similar layer systems reported in [43]. XRD signals for samples of Ru/Si at 300 K, Ru/Si after an RTP anneal at 600°C for 10 min, and for Ru/SiO_2/Si after an anneal at 600°C for 10 min are also shown in **Figure 8a**. It is seen that a Ru_2Si_3 crystallization peak appears as a result of the 600°C anneal. An XRD signal (more of a shoulder than a peak) of ruthenium silicide Ru_2Si_3 at 44.7° is observed next to a broad Si wafer peak only after 900°C anneal as shown in **Figure 8a**. This is in agreement with observations of Ru_2Si_3 XRD detection made in ref. [44]. We observe this peak on the following samples: Ru/Si, Ru/SiO_2/Si, Ru/Ti/SiO_2/Si, and Cu/Ru/Ti/SiO_2/Si. The strongest Ru_2Si_3 signal can be seen for the Ru/Si and Ru/SiO_2 samples. From comparison of **Figure 8b** with **Figure 8a** it, can be seen that the XRD signal at 44.7° is absent for the anneal at 600° C and also for the unannealed sample. The Ru_2Si_3 shoulder is either not there or very weak, as seen in **Figure 8b**. From **Figure 8b** it is also seen that the Ru_2Si_3 peak at 41.7°

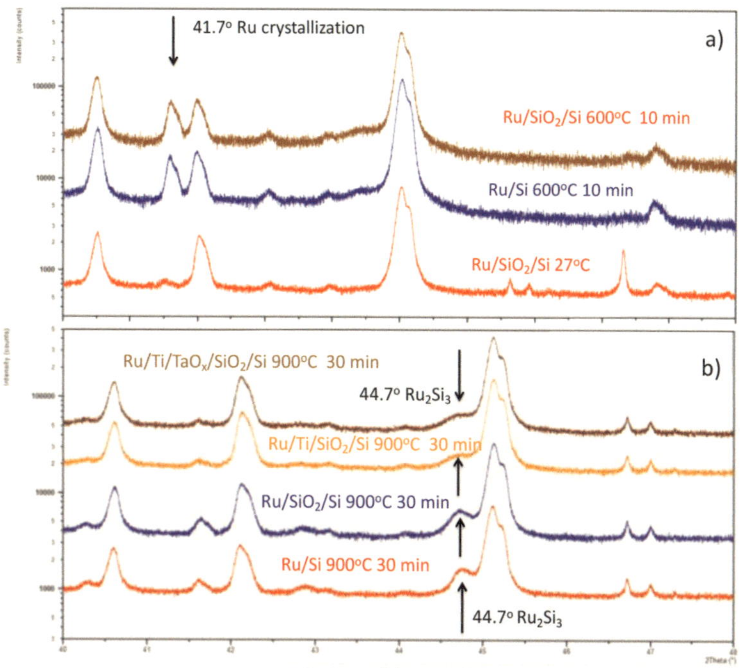

Figure 8.
XRD spectra of (a) three-layer structures: Ru/SiO_2/Si as deposited, Ru/Si and Ru/SiO_2/Si after a 600°C 10 min anneal. The Ru crystallization peak has been also observed at other layer structures with a Ru layer only after a 600°C anneal. The peak is very weak at room temperature and disappears after an anneal at 900°C for 30 min. (b) Four-layer structures annealed at 900°C for 30 min: Ru/Si, Ru/SiO_2/Si, Ru/Ti/SiO_2/Si and Ru/Ti/TaO_x/SiO_2/Si. On all structures a Ru_2Si_3 shoulder can be observed at 44.7°C next to a large signal 45.12°C which stems from the Si wafer as verified by XRD spectra on Si and SiO_2/Si structures [35] © The Electrochemical Society reproduced by permission of IOP Publishing. Al-Mamun M, King, S, Orlowski M. "Thermal and Chemical Integrity of Ru Electrode in Cu/TaOx/Ru ReRAM Memory Cell", ECS J. Sol. St. Technol.2019; 8: N220; DOI: 10.1149/2.0121912jss.

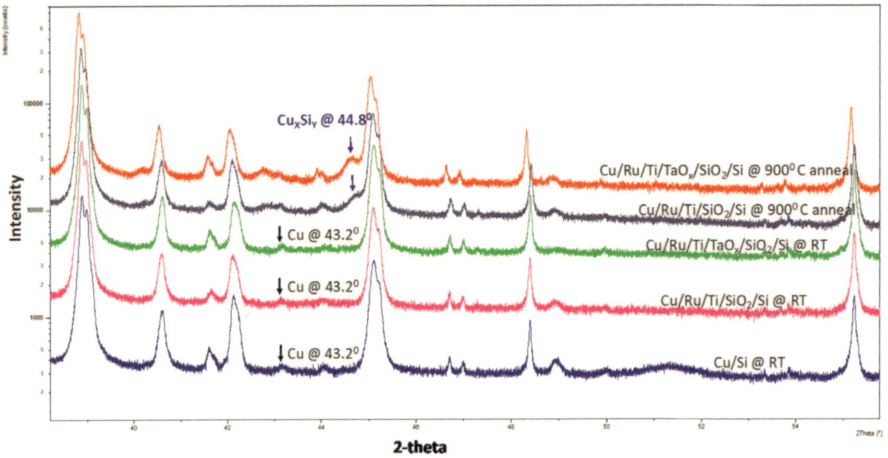

Figure 9.
*XRD spectra of the same layer structures as in **Figure 8b** but capped by Cu layer. A Cu silicide peak can be observed at 44.8° only after the anneal at 900°C [35].*

disappears at 900°C, again in agreement with observations made in [43]. In layer structures of **Figure 8b** that have been capped with an additional Cu layer, the XRD spectra show formation of copper silicide at 44.8° only at 900°C as shown in **Figure 9**. Thus, at high temperatures Cu and Si may diffuse through Ru/Ti and Ru/Ti/TaO_x layers to undergo a copper silicide reaction.

In summary, the XRD studies have revealed formation of Ru grains, formation of two kinds of Ru silicides, Ru_2Si_3 and Ru_2Si_3, and that insertion of the TaO_x layer between SiO_2 and Ti acts as a diffusion barrier for Si and suppresses the Ru_2Si_3 reaction. The XRD study corroborates not only the difference in electrical performance between Pt and Ru device but also among the Ru devices embedded on different substrates. Considering these results, it appears that while the inertness properties of Pt and Ru are similar up to about 200°C, at elevated temperatures, Pt is by far more inert than Ru ensuring superior switching performance of the Cu/TaO_x/Pt over the Cu/TaO_x/Ru device.

9. Conclusions

Against the backdrop of analogously characterized Cu/TaO_x/Pt and Cu/TaO_x/Co devices, we have used electrical properties of a Cu/TaO_x/Ru ReRAM cell, to elucidate material properties of 50 nm ruthenium films. The circumstance that an electric switching operation of a ReRAM cell allows to raise temperature of the Cu filament inside a ReRAM memory cell from 27°C up to ca. 1100°C, in a controlled way [45], turns the Cu/TaO_x/Ru device into a nano-laboratory for investigation of the material properties of Ru at different temperatures. The present electric analysis of the devices has been complemented by extensive material studies of relevant layer systems annealed at various temperatures using XRD, and by studies of the surface roughness of free and passivated metal surfaces, both unannealed as well as annealed at elevated temperatures, using AFM. From these studies, a consistent picture emerges allowing a compelling explanation of all differences of the ReRAM cell's electric characteristics in terms of the Ru material properties.

Cu/TaO_x/Ru devices compared to Cu/TaO_x/Pt devices have higher forming, set, and reset voltages. The higher V_{form} and V_{set} voltages can be partly or entirely attributed to the work function difference between Pt and Ru of 1.6 eV and partly to the significant difference in surface roughness between Pt and Ru. The higher V_{res} voltage for Ru device than for Pt or Co devices is a consequence of much higher thermal conductivity of Ru than that of Pt and Co. It was also found that while Pt devices can be switched more than hundred times, the number of switching cycles of Ru devices is very limited. The performance deterioration of Ru devices is particularly conspicuous when the cell is exposed to high Joule's heat dissipation during the set and reset switching cycles. At low Joule's heat dissipation, the switching performance of both Ru and Pt devices is initially comparable. The Joule's heat dissipation is, thus, a major factor determining the endurance properties of the device as it triggers degradation of the inertness properties of Ru caused by grain crystallization as well as silicide formation. The degraded integrity of the inert electrode appears to have a major impact on the Cu diffusional fluxes which, in turn, determine the geometrical shape of the Cu conductive filament. Additional embedment experiments have corroborated these conclusions. In an embedment experiment, an identical device proper is embedded in different ways on the Si wafer. It has been demonstrated that the electrical switching properties of nominally the same resistive switching device Cu/TaO_x/Ru, when embedded differently, Cu/TaO_x/Ru/Ti/SiO_2/Si vs. Cu/TaO_x/Ru/Ti/TaO_x/SiO_2/Si, differ substantially. This is the result of chemical and structural interactions of the device proper with its immediate environment. The chemical and structural changes of Ru films are brought about by the nanometer dimension of the layers involved and triggered by Joule's heat deposition during the switching of the ReRAM cell. This finding points to a broader and critical impact of the device's embedment on its structural, material integrity, and, eventually, its electrical reliability and endurance properties. The insertion of a TaO_x layer between the SiO_2 and Ti glue layer suppresses the Ru crystallization and silicide reaction somewhat and, eo ipso, improves the electrical performance and endurance of the Ru device on the Ti/TaO_x/SiO_2/Si substrate as compared to the Ru device manufactured on the Ti/SiO_2/Si substrate. Embedment structures with different glue layers (Cr vs. Ti) have corroborated the role of thermal conductivity in determining the actual value of V_{res}. The surface roughness studies have shown a strong impact of surface roughness on V_{form}, evidenced by the three devices. Furthermore, it turned out that deposition of a sacrificial dielectric layer followed by an anneal leads to a drastic reduction of surface roughness of a metal.

The advantages and disadvantages of Ru over Pt devices are summed up in **Table 4**. The most important disadvantage of the Ru device is its limited switching cycling ability. This could be perhaps addressed in future work by minimizing Joule's

Inert electrode	Threshold voltages	retention	Endurance cycling ability	BEOL integrability
Ru	high (−)	(+)	<14 (−)	(++)
Pt	low (+)	(++)	>200 (++)	(−)

Table 4.
Figure of merit comparison between Ru and Pt ReRAM device. (+) and (−) denote advantages and disadvantages, respectively.

heat exposure during the switching operations. The use of ruthenium in ReRAM cells appears to be viable if the internal temperature of the cell could be limited to temperature not exceeding ca 350°C, which puts a severe constraint on how the cells are being programmed and erased. Up to 440°C, Ru exhibits similar stopping power and inertness properties as Pt. Hence, the deployment of Ru in BEOL, where the temperature during a circuit operation does not exceed 440°C [46], appears to be viable. However, at elevated temperatures above ca 400°C, the inertness properties of Pt are vastly superior to those of Ru.

Author details

Amrita Chakraborty, Mohammad Al-Mamun and Marius Orlowski*
Bradley Department of Electrical and Computer Engineering, Virginia Polytechnic Institute and State University (Virginia Tech), Blacksburg, VA, USA

*Address all correspondence to: m.orlowski@vt.edu

References

[1] Liu T, Kang K, El-Helw S, Potnis T, Orlowski M. Physics of the voltage constant in multilevel switching of conductive bridge resistive memory. Japanese Journal of Applied Physics. 2013;**52**:084202. DOI: 10.7567/JJAP.52.084202

[2] Al-Mamun M, Orlowski M. Challenges to implement resistive memory cells in the CMOS BEOL. ECS Transactions. 2017;**80**(6):13-23

[3] Benasconi R, Magagnin L. Review-ruthenium as diffusion barrier layer in electronic interconnects. Journal of ECS. 2019;**166**(1):D3219-D3225

[4] Wen LG, Adelmann C, Pedreira OV, Dutta S, Popovici M. Ruthenium metallization for advanced interconnects. IEEE Internet Interconnect Technique. 2016;**2016**:34-36

[5] Zahedmanesh H, Gonzales VV, Tokei Z. Nano-ridge bending during conformal ruthenium metallization. ACS Applied Nano Materials. 2021;**4**:5643-5648

[6] Tada M, Okamoto K, Sakamoto T, Miyamura M, Banno N, Hada H. Polymer solid-electrolyte switch embedded on CMOS for nonvolatile crossbar switch. IEEE Transactions on Electrical Development. 2011;**58**:4398

[7] The International Technology Roadmap For Semiconductors. 2015. Available from: http://www.semiconductors.org/clientuploads/Research_Technology/ITRS/2015

[8] Liu TY, Yan TH, Scheuerlein R, Chen Y, Lee JK, Balakrishnan G. A 130.7mm^2 2-layer 32Gb ReRAM memory device in 24nm technology. In: 2013 IEEE Sol.-S. Circ. Conf. Digest. San Francisco. 2013. pp. 210-211. DOI: 10.1109/ISSCC.2013.6487703

[9] Orlowski M, Ghosh G, Verma A. Method to manufacture highly conductive vias & PROM memory cells by application of electric pulses. US Patent Application Publication 2019/0058120 A1, Feb. 21, 2019

[10] Yang JJ, Strukov DB, Stewart DR. Memristive devices for computing. Nature Nanotechnology. 2013;**8**:13

[11] Strukov B, Snider DR, Stewart R, Williams RS. The fourth circuit element. Nature. 2008;**453**:80

[12] Chua LO, Kang SM. Memristive devices and systems. Proceedings of the IEEE. 1976;**64**(2):209-223. DOI: 10.1109/PROC.1976.10092

[13] An H, Al-Mamun MS, Orlowski MK, Yi Y. Learning accuracy analysis of memristor-based nonlinear computing module on long short-term memory. In: Proceedings of the International Conference on Neuromorphic Systems, Knoxville, Tennessee. ACM: New York, NY; 2018

[14] Beigi MV, Memik G. Thermal-aware optimizations of ReRAM-based neuromorphic computing systems. In: Proc. 55th Annual Des. Autom. Conf. San Francisco, 24-28 June 2018. p. 324

[15] Azghadi MF, Chen YC, Eshraghian JK, Chen J, Lin CY, Amirsoleimani A, et al. Complementary metal oxide semiconductor and Memristive hardware for neuromorphic computing. Advanced Intellectual System. 2020;**2**:1900189

[16] Waser R, Dittmann R, Staikov G, Szot K. Redox-based resistive switching

memories – Nanoionic mechanisms, prospects, and challenges. Advanced Materials. 2009;**21**:2632

[17] Kaeriyama S, Sakamoto T, Sunamura H, Mizuno M, Kawaura H, Hasegawa T, et al. A nonvolatile programmable solid-electrolyte nanometer switch. IEEE Journal of Solid-State Circuits. 2005;**40**:168

[18] Kozicki MN, Park M, M. Mitkova M. Nanoscale memory elements based on solid-state electrolytes. IEEE Transactions on Nanotechnology. 2005;**4**:331

[19] Fowler BW, Chang YF, Zhou F, Wang Y, Chen PY, Xue F, et al. Electroforming and resistive switching in silicon dioxide resistive memory. RSC Advances. 2015;**5**:21215-21236

[20] Xiao B, Gu T, Tada T, Watanabe S. Conduction paths in Cu/a-Ta2O5/Pt atomic switch: First-principles studies. Journal of Applied Physics. 2014;**115**: 034503

[21] Ghosh G, Kang Y, King SW, Orlowski M. Role of CMOS Back-end metals as active electrodes for resistive switching in ReRAM cells. ECS Journal of Solid State Science and Technology. 2017;**6**(1):N1-N9

[22] Calderoni A, Sills S, Cardon C, Faraoni E, Ramaswamy N. Engineering ReRAM for high-density applications. Microelectronic Engineering. 2015; **137**:145

[23] Ohno T, Samukawa S. Resistive switching in a few nanometers thick tantalum oxide film formed by a metal oxidation. Applied Physics Letters. 2015; **106**:173110

[24] Kurnia F, Jung C, Lee B, Liu C. The evolution of conducting filaments in forming-free resistive switching Pt/TaOx/Pt structures. Applied Physics Letters. 2015;**107**:073501

[25] Kawahara A, Azuma R, Ikeda Y, Kawai K, Katoh Y, Tanabe K, et al. An 8Mb multi-layered cross-point ReRAM macro with 443MB/s write throughput. IEEE International Solid-State Circuits Conference. 2012;**2012**:432-434. DOI: 10.1109/ISSCC.2012.6177078

[26] Rullan J, Ishizaka T, Cerio F, Mizuno S, Yang CC. Low-resistance wiring and 2xnm void-free fill with CVD ruthenium liner and direct seed copper. Solid State Technology. 2016;**2016**:432

[27] Standaer T, Beique G, Chen H-C, Chen S-T, Hamieh B, Lee J. BEOL process integration for the 7nm technology node. IEEE Interconnect Technical Conference. 2016;**2016**:2-4. DOI: 10.1109/IITC-AMC.2016.7507636

[28] Masalski TB. Cu-Ru Phase Diagram, in Binary Alloy Phase Diagrams, Editors Masalski, Okamoto. Materials Park: American Society of Metals; 1990. p. 1467

[29] Kang Y, Verma M, Liu T, Orlowski M. Formation and rupture of multiple conductive filaments in Cu/ TaO_x/Pt device. ECS Solid State Letters. 2012;**1**:48-50

[30] Liu T, Kang Y, Verma M, Orlowski M. I–V characteristics of antiparallel resistive switches observed in a single Cu/TaO_x/Pt Cell. IEEE Electrode Deviation Letters. 2012;**33**:429

[31] Liu T, Verma M, Kang Y, Orlowski M. Coexistence of Bipolar and Unipolar Switching of Cu and Oxygen Vacancy Nanofilaments in Cu/TaOx/Pt Resistive Devices. ECS Solid State Letters. 2012;**1**:Q11

[32] Ghosh G, Orlowski M. Write and erase threshold voltage interdependence

in resistive switching memory cells. Transactions on IEEE Electronic Devices. 2015;**62**:2850

[33] Ghosh G, Orlowski M. Correlation between set and reset voltages in resistive RAM cells. Current Applied Physics. 2015;**15**:1124

[34] Chakraborty A, Al-Mamun M, Orlowski M. Impact of surface roughness and material properties of inert electrodes on the threshold voltages and their distributions of ReRAM memory cells. ECS Journal of Solid State Science Technology. 2022;**11**:104007

[35] Al-Mamun M, King S, Orlowski M. Thermal and chemical integrity of Ru electrode in Cu/TaOx/Ru ReRAM memory cell. ECS Journal of Solid State Technology. 2019;**8**:N220

[36] Ali R, Fan Y, King SW, Orlowski M. Modeling and simulation of Cu diffusion and drift in porous CMOS backend dielectrics. APL Materials. 2018;**6**: 066101

[37] Al-Mamun S, Orlowski M. Performance degradation of nanofilament switching due to Joule. Heat Dissipation Electronics. 2020;**9**(1):127

[38] Al-Mamun M, Chakraborty A, Orlowski M. Methodology for Mitigation of ReRAM Reliability Issues Caused by Thermal Cross-Talk. 2022 submitted for publication

[39] Eyring C, Milikan R. Field currents from points. Physical Review. 1928;**31**: 900

[40] Kim HJ, Kaplan KE, Schindler P, Xu S, Winterkorn MM, Heinz DB, et al. Electrical properties of ultrathin platinum films by plasma enhanced atomic layer deposition. ACS Applied Materials & Interfaces. 2019;**11**:9594

[41] Simpson DL, Crosswell RT, Reisman A, Temple D, Williams CK. Planarization processes and applications - I. Undoped GeO2-SiO2 glasses. Journal of Electro-Chemical Society. 1999;**146**(10):3860

[42] McKinley KA, Sandler NP. Tantalum pentoxide for advanced DRAM applications. Materials Research Society Symposium Proceedings. 1997;**446**: 299-307

[43] Peterson CS, Baglin JE, Dempsy JJ, La Placa S. Silicides of ruthenium and osmium: Thin film reactions, diffusion, nucleation, and stability. Journal of Applied Physics. 1982;**53**:4886

[44] Arunagiri TN, Zhang Y, Chyan O, El-Bouanani M, Kim MJ, Chen KH, et al. 5nm ruthenium thin film as a directly plateable copper diffusion barrier. Applied Physics Letters. 2005;**86**:083104

[45] Al-Mamun M, Orlowski M. Electron Tunneling between vibrating atoms in a copper nano-filament. Scientific Reports. 2021;**11**:7413

[46] Jenkins M, Austin DZ, Conley JF, Ghosh G, Orlowski M, King SW. Review—Beyond the highs and lows: A perspective on the future of dielectrics research for Nanoelectronic devices. ECS Journal of Solid State Science & Technology. 2019;**8**:N159

Chapter 4

An Insight towards the Design of a Ruthenium-Containing Biomaterial

Duduzile Nkomo, Maje Phasha and Hein Moller

Abstract

Ruthenium (Ru) is one of the platinum group metals (PGMs). These metals belong to the transition metals group of the periodic table. They have excellent properties such as high melting point and are inert with variety of substances, thus also called noble metals. Currently, Ru is the cheapest of the PGMs, thus it is readily available compared to other PGMs. Recently, incorporating PGMs in shape memory alloys (SMAs) has been extensively explored, with titanium-nickel (TiNi) used as a benchmark material. TiRu is amongst the compounds that are currently explored for various potential applications. This compound has an ordered B2 (CsCl-type) crystal structure. It is hard and brittle, thus some shape memory (SM) properties are difficult to induce in this compound. However, due to Ru possessing some good biomedical properties such as biocompatibility, corrosion resistance, improved radiopacity and ultra-low magnetic susceptibility for MRI diagnostics, the mechanical properties of TiRu must be improved for biomedical applications. Since niobium (Nb) is known to be biocompatible and is usually studied in biomedical alloys, a systematic substitution of Ti with niobium (Nb) was performed in an effort to reduce the stiffness (Young's modulus). This chapter gives an insight on the structural and mechanical properties of biocompatible Ru-rich alloy compositions.

Keywords: ruthenium, materials design, biomedical alloys, elastic properties, SMAs

1. Introduction

The equilibrium phases of the Ti-Ru system are the disordered bcc (βTi) solid solution, the ordered cubic compound (B2) TiRu and the close-packed hexagonal solid solution. **Figure 1** shows the experimentally obtained phase diagram of the Ti-Ru system. It can be observed that the β to α transformation temperature decreases from 882°C with an increase in Ru content up to 9 atomic percent (at.%), after which the β phase is stabilized. The maximum solubility of Ru in βTi is about 25 at.% at the peritectic temperature of 1575°C. On the other hand, the B2 TiRu compound is observed at composition range of 43–53 at.% Ru. TiRu congruently melts at 2130°C, below which the B2 phase remains stable to low temperatures [1].

Due to lower cost compared to other PGMs, Ru addition has been explored in different alloy systems (β and B2 type alloys) in order to improve various properties [2–5]. In βTiNb alloys which hold much promise in the biomedical field because of

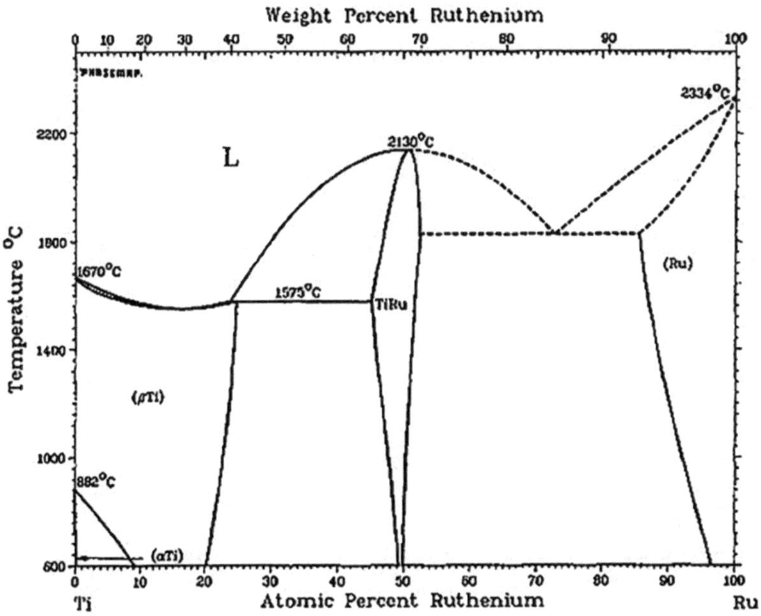

Figure 1.
Ti-Ru system phase diagram [1].

intrinsic properties such as low modulus, biocompatibility and shape memory properties, such as superelasticity (SE), Ru is expected to increase strength, corrosion resistance, improve radiopacity and decrease magnetic susceptibility. Shape memory effect (SME) typically encompasses the phase transformation between two phases. For β-alloys, the reversion occurs between the high temperature β-phase and low temperature orthorhombic martensite α′′-phase at a critical alloying compositions, whilst for CsCl-type intermetallic compounds-like alloys the reversion is typically between the high-symmetry B2 and low-symmetry B19′/B19/B33/$L1_0$ phase [6]. It is this phase transformation that governs the functionality of SMAs. SE is important for biomedical applications such as orthodontic arc wires, bone plates and stents because it allows for shape recovery after deformation [4, 7]. However, recoverable strain have been found to be small in βTiNb alloys. Kim and Miyazaki [8] observed that alloys containing more that 25 at.% Nb had a recoverable strain of about 2.5%, which is much lower than that of TiNi alloys, which is ~8%.

In order to improve SE at room temperature, addition of ternary elements to TiNb has been considered. The shape memory properties of ternary Ti-Nb-X (X = Si, Mo, O, N, Zr, Pd, Al, Ag, Pt, etc.) alloys have been extensively investigated [9–18].

However, there are only limited studies that considered Ru as a ternary element in TiNb alloys [4, 19]. This remains so despite several improvements that were observed due to addition of Ru in Ti-20Nb-xRu (x = 0, 0.5, 1.0, 1.5 at.%). The yield strength ranged from 560 to 900 MPa and the Young's modulus ranged from 60 to 100 GPa, with increasing Ru content. Furthermore, it was observed that Ru acted as a β-phase stabilizer and thus showed a potential to promote the existence of SE at room temperature [4]. The occurrence of SE in shape memory alloys is detailed elsewhere [20]. Although, according to authors' knowledge the SE of Ti-20Nb-xRu (x = 0, 0.5, 1.0, 1.5 at.%) has not been published, it can be deduced that 20 at.% Nb could still yield lower transformation strain as it is reported that this strain can only be improved at Nb content less than 15 at.% [7, 8].

Previous studies [4, 19] have shown that Ru also improves properties such as corrosion resistance and lower magnetic susceptibility. Corrosion resistance is an important characteristic in biomedical alloys because corrosion results in harmful ionic release into the body and poor osseointergration due to metal loss. The effect of Ru addition in TiNb alloys has been reported from investigations of Ti-20Nb-xRu by Biesiekierski et al. [19] and it was revealed that Ru-containing alloys were more corrosion resistant compared to Ru-free alloys. Furthermore, it has been reported that Ru can enhance the corrosion resistance of biomedical Ti-based alloys by several orders of magnitude with even a 0.1 at.% addition in particular alloy systems [19, 21].

Moreover, it has been reported that most conventional metallic biomaterials with high magnetic susceptibility have adverse effects on the Magnetic Resonance Imaging (MRI) and diagnostics as they would become magnetized in the intense magnetic field of the MRI. This causes heat generation in the biomedical materials and therapeutic devices, which leads to displacement of biomaterials and therapeutic devices, and artifacts on the MRI. Such artifacts can distort the authentic bio-imaging of the human organs and tissues around the implant. Materials and devices with an ultra-low magnetic susceptibility are required for surgery and diagnostics performed under MRI. The βZrRu alloys with ultra-low magnetic susceptibility have been developed for biomedical and therapeutic devices. Here, adding Ru to Zr reduced magnetic susceptibility, as shown in **Figure 2** [2].

In ordered B2 systems, the effect of Ru addition has been reported in B2 $Ti_{50}Ni_{50-x}Ru_x$, $Ti_{50}Pt_{50-x}Ru_x$, $Zr_{50}Pd_{50-x}Ru_x$ potential SMAs [3, 5, 22]. The phase stability, transformation temperature and elastic properties has been studied in the B2 $Zr_{50}Pd_{50-x}Ru_x$ (x = 0, 6.25, 12.50, 18.75) alloy compositions using computational methods [5] . The transformation temperature of $Zr_{50}Pd_{50-x}Ru_x$ alloys was observed to be sensitive to the Ru content, where B2 phase became more stable with Ru

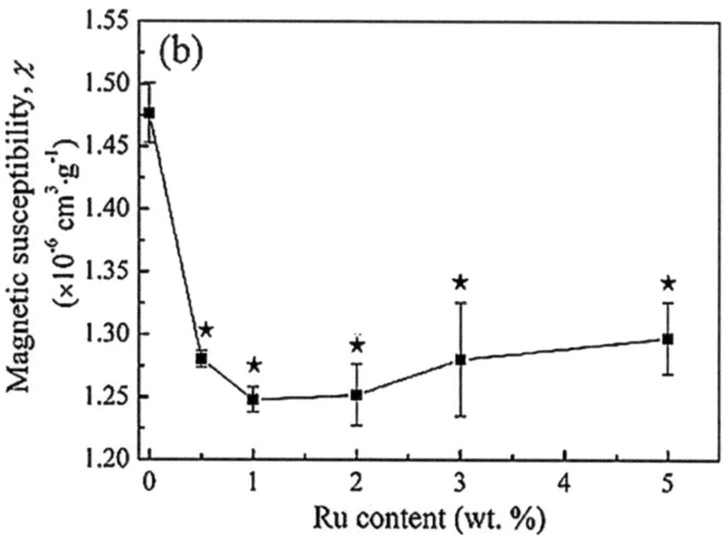

Figure 2.
The magnetic susceptibility of pure Zr and βZrRu alloys [2].

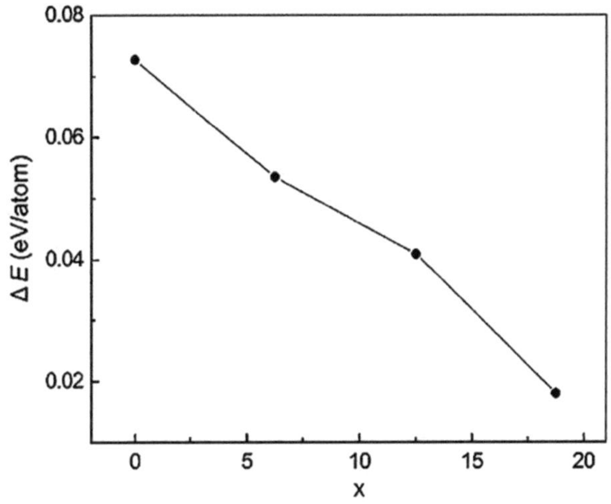

Figure 3.
The total energy difference (ΔE) between austenite and martensite phases of $Zr_{50}Pd_{50-x}Ru_x$ *alloys [5].*

addition. This was shown by calculation of total energy difference (ΔE) between the austenite B2 and orthorhombic martensite phase. It was observed that the total ΔE decreased with increasing Ru content as shown in **Figure 3**, meaning that the transformation temperature of $Zr_{50}Pd_{50-x}Ru_x$ alloys decreased with increasing Ru content.

Also, a preceding experimental study on similar alloys with even higher Ru contents showed that transformation temperature decreased from about 620°C at x = 0 binary alloy to around 0°C for x = 35 alloy [5, 23], indicating the B2 phase-stabilizing effects of Ru. These findings were similar to work done by Tsuji et al. [3], where systematic substitution of Ni with Ru in TiNi stabilized the B2 phase to the point where martensitic phase transformation completely diminished.

In B2 $Ti_{50}Pt_{50-x}Ru_x$, the partial substitution (5 at.%) of Pt with Ru decreased the transformation temperature from martensitic finish (M_f) =1285 K for TiPt to M_f = 1129 K for $Ti_{50}Pt_{45}Ru_5$, further proving the B2-stabilizing effect of Ru. The compressive strength obtained at about 1129 K was slightly improved from ~500 MPa for TiPt to ~700 MPa for $Ti_{50}Pt_{45}Ru_5$. This means that partial substitution of Pt with Ru increased the critical stress for slip deformation. Furthermore, the recovery ratio due to shape memory effect was improved from ~11% for TiPt to ~45% for $Ti_{50}Pt_{45}Ru_5$ [22].

Furthermore, the B2 NbRu and TaRu have been studied as potential SMAs for high temperature applications [24]. Both compounds exhibit a B2 phase at higher temperatures and starts transforming to tetragonal ($L1_0$) martensitic phase below ~1158 K for NbRu and ~1403 K for TaRu. The shape recovery of NbRu and TaRu was reported to be 73% and 80%, respectively [24, 25], which is better than ones for B2 TiPt [22] and TiRu [3] at higher temperatures.

Although Ru is expensive compared to other transition metals considered for biomaterials, it is extremely promising for potential biomedical applications because it exhibits low ionic cytotoxicity *in vitro*, has excellent biocompatibility *in vivo* and no evidence of mutagenicity or carcinogenicity [21]. Moreover, it has good osteocompatibility that can exceed that of conventional pure Ti and Ti-based biomaterials [2].

At this point, the advantages of Ru addition in both β and B2-type alloys have been outlined. In view of the biomedical properties improved by Ru, it is worth a while to explore alloys with high Ru content, which means studying B2 alloys that form at high Ru compositions. Thus, the current study starts with B2 TiRu alloy that is stable to below ~874 K, a factor that often makes it difficult to obtain the SM properties. Since SE is also dependent on the elastic moduli of material, the underlying mechanical properties of B2-type alloys must be improved for consideration in biomedical applications. Specifically, in this case the aim is to improve the mechanical properties of TiRu alloy for biomedical applications. Therefore, this chapter outlines the underlying mechanical properties induced by addition of Nb in TiRu in order to give an insight towards the design of a biomaterial with possible SME. The prediction of elastic properties has been performed for the B2 Ti-Nb-Ru alloys using the DFT method to guide the reader towards the design and development of Ru-rich biomaterials.

2. Computational details

To the authors' knowledge, there is currently no experimental data available on the alloy system investigated in this chapter. Therefore, the authors have used similar approaches to that employed by other authors [5, 26–28] in order to have basis for the current research. The calculations in this study were performed on a unit cell and 2 × 2 × 2 supercell of B2-type crystal structure with space group #221(Pm-3m) consisting of 2 and 16 atoms, respectively, using DFT based CASTEP code embedded in Materials Studio 2020 software package [29]. Robust Vanderbilt ultrasoft pseudopotentials [30] were used to describe the ion-electron interaction within the generalized gradient approximation (GGA) [31] of Perdew-Burke-Ernzerhof (PBE) [32]. A plane wave energy cutoff of 700 eV and k-point set of 16x16x16 and 8x8x8 for unit cell and supercell, respectively, were sufficient to converge the total energy of the considered systems [33]. The effect of high Ru content and then substituting Ti atoms with Nb on the structure and elastic properties was determined. The starting composition was a binary Ti_8Ru_8, representing 50 at.% Ru and 0 at.% Nb. Then, the ternary compositions considered were of stoichiometry, $Ti_7Nb_1Ru_8$, $Ti_6Nb_2Ru_8$, $Ti_5Nb_3Ru_8$ and $Ti_4Nb_4Ru_8$, representing 6.25, 12.50, 18.75 and 25.0 at.% of Nb, as shown in **Figure 4**. All the ground-state structures were optimized using the Brayden-Fletcher-Goldfarb-Shanno (BFGS) minimization scheme. The convergence criterion

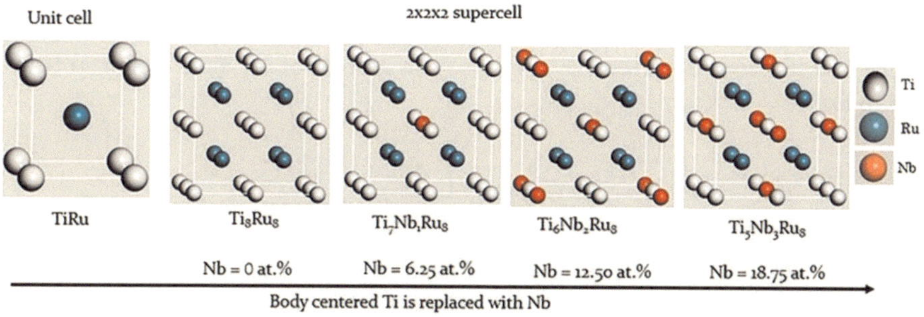

Figure 4.
Crystallographic structures of B2 $Ti_{(50-x)}Nb_xRu_{50}$ alloys.

$Ti_{50}Ru_{50}$	3.078
$Ti_{43.75}Nb_{6.25}Ru_{50}$	3.092
$Ti_{37.50}Nb_{12.50}Ru_{50}$	3.104
$Ti_{31.50}Nb_{18.50}Ru_{50}$	3.115
$Ti_{25}Nb_{25}Ru_{50}$	3.280

Table 1.
Theoretical lattice parameters (Å) of equilibrium cubic crystals (a_o) of B2 $Ti_{(50-x)}Nb_xRu_{50}$ alloy compositions.

of less than 1×10^{-5} eV/atom, the maximum residual forces of 0.03 eV/Å, maximum residual bulk stress of 0.05 GPa and maximum atomic displacement of 1×10^{-3} Å were utilized.

The theoretical lattice parameters for cubic crystals of B2 $Ti_{(50-x)}Nb_xRu_{50}$ alloys were calculated and are presented in **Table 1**. The experimental lattice parameter for $Ti_{50}Ru_{50}$ is 3.067 Å [1, 3], which agrees with the theoretical value within the percentage error of 0.36%. The experimental values for ternary alloys are not yet available in literature. The atomic radii of Ru, Nb and Ti are 1.34, 1.47 and 1.46 Å, respectively [34]. Given that Nb and Ti have almost similar atomic size, it is noted that lattice constant (a_o) increased with increasing Nb content, owing to high coefficient of thermal expansion of Nb. This has been observed in the work conducted by Bonisch et al. [11] where the lattice parameters showed dependence on Nb content in TiNb. The linear thermal expansion of the Nb-rich orthorhombic martensite phase was found to be larger along the lattice parameter *a* than lattice parameters *b* and *c*. The *a* expanded at a rate of 163×10^{-6}°C^{-1} at temperatures below 500 K, indicating the high thermal expansion of 9.2×10^{-6}°C^{-1} for Nb compared to Ti which is 8.4×10^{-6}°C^{-1} [35].

3. Mechanical properties

3.1 Mechanical stability

The mechanical properties of materials could be estimated to some extent by analyzing the elastic constants. However, very accurate calculation of elastic constants is essential for gaining meaningful insight into the mechanical stability and elastic properties of materials. Elastic constants of a mechanically stable phase should meet the requirements of mechanical stability based on the Born Standard. The elastic behavior of a lattice is described by its matrix of second order elastic constants as shown in Eq. (1) [36].

$$C_{ij} = \frac{1}{V_0}\left(\frac{\partial^2 E}{\partial_{\varepsilon_i}\partial_{\varepsilon_j}}\right) \tag{1}$$

Where: E is the energy of the crystal, V_0 is the equilibrium volume and ε is strain. The elastic matrix has a size of 6×6 and is symmetric, thus composed of 21 independent components. A crystalline structure is stable in the absence of external load and in the harmonic approximation only if all its phonon modes have positive frequencies for all wave vectors. The elastic energy E is given by Eq. (2), and E_0 is the minimum energy of a relaxed crystal structure. The equation gives the energy of the crystal from the arbitrary deformation by an infinitesimal strain [36].

$$E = E_0 + \frac{1}{2} V_0 \sum_{i,j=1}^{6} C_{ij}\varepsilon_i\varepsilon_j + O\left(\varepsilon^3\right) \tag{2}$$

For cubic crystals, the minimum energy conformation requires about four simulations. The derivation of coefficients requires an additional three or four simulations per independent coefficient. Fifteen energy calculations are required from which three independent constants C_{11}, C_{44} and C_{12} are derived for a cubic crystal. The mechanical stability criteria for cubic system is explained in [37] and is given as follows:

$$C_{11} - C_{12} > 0; C_{44} > 0; C_{11} + 2C_{12} > 0 \tag{3}$$

Using the calculated equilibrium lattice constants, the single-crystal elastic constants for B2 $Ti_{50-x}Nb_xRu_{50}$ lattice structures were calculated in the current study in order to determine the effects of systematically substituting Ti with Nb on the stability and elastic properties of the B2 phase. There is currently no experimental values available for comparison for these ternary alloy compositions, only the Vickers hardness (VH) of TiRu has been found to be around 400 GPa [3]. Based on these three independent single crystal elastic constants of a cubic crystal, the tetragonal shear modulus C' is determined using the following expression:

$$C' = \frac{C_{11} - C_{12}}{2} \tag{4}$$

The tetragonal shear modulus has been found to be significant in quantifying the mechanical stability. A positive C' indicates the mechanical stability of a crystal [38]. **Table 2** shows the calculated elastic constants C_{ij} and C'.

The calculated elastic constants are all positive, which indicates that the mechanical stability criteria in Eq. (3) is satisfied. Also the tetragonal shear modulus C' is positive, which further indicates the stability of B2 phase. Since these calculations are carried out at 0 K temperature, this therefore implies that B2 phase is stable even below room temperature. **Figure 5** shows the graphical presentation of the elasticity data, and it can be observed that C' decreases with increasing Nb concentration (i.e.,

x (at.% Nb)	Elastic constants (GPa)			
	C_{11}	C_{44}	C_{12}	C'
0.00	421	88	114	154
6.25	401	86	119	141
12.50	396	87	133	132
18.75	342	84	160	91
25.00	297	84	195	51

Table 2.
Theoretical single-crystal elastic constants of B2 $Ti_{50-x}Nb_xRu_{50}$ alloys.

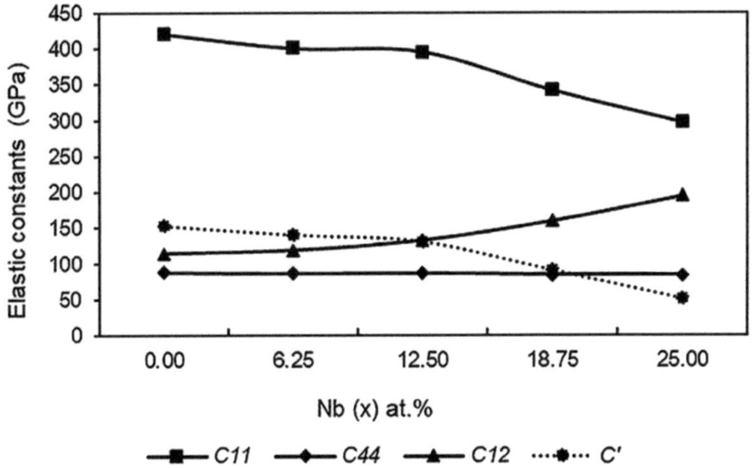

Figure 5.
Elastic constants of $Ti_{50\text{-}x}Nb_xRu_{50}$ alloys as a function of Nb content.

with decreasing Ti concentration). While elastic constant C_{44} remains almost constant, the C' reduces slightly for Nb compositions up to 12.50 at.% (change <15 GPa), mainly due to high mechanical stability of B2 TiRu and its strong bonding character, hence the small increment in Nb does not create a significant change. However, at 18.75 at.% Nb and above, the decrease in C' starts to be more pronounced. This decrease is attributed to higher content of more ductile Nb in the place of less ductile Ti. As indicated, presence of Nb results in the expansion of the average volume per atom in Ti-Nb-Ru, which means a decrease of the average bond strength and consequently reduces the ability of the alloy to resist the tetragonal shear. Furthermore, the crystal structure theory of transition metals states that solids with increasing d-occupation number increases its metallicity, a factor that is linked to ductility [39].

The stability of B2 TiRu phase has been experimentally studied by Murray [1] as indicated by the phase diagram in **Figure 1** in the introductory chapter. It was shown that, the B2 phase undergoes congruent melting at about 2130°C. Below this temperature the B2 phase remains stable to low temperatures for compositional ranges of 45 to 52 at.% Ru. The obvious observation here is that with higher Ru concentration the crystal structure moves from being disordered β to ordered B2. At this point, it was beneficial to study the mechanical properties of the ordered B2 structures in order to further gain an insight on the underlying contribution of Ru. As indicated, the introduction of Nb has proven to reduce the stability of B2 phase as shown by a substantial decrease in C', thus we also further investigate the impact of Nb addition on other mechanical properties.

3.2 Bulk modulus, *B*

The bulk modulus is a measure of resistance to volume change or to compression deformation by applied pressure. Based on the three independent single-crystal elastic constants of a cubic crystal, the bulk modulus B is expressed as follows [40]:

$$B = \frac{C_{11} + 2C_{12}}{3} \tag{5}$$

Figure 6 shows the graph of B for B2 $Ti_{50-x}Nb_xRu_{50}$ alloys. The B varies with increasing Nb content. Previous studies have suggested that crystals with larger lattice constants possess lower average bond strength and thus can be easily compressed compared to those with smaller lattice constants [39]. As such, an anomalous behavior is observed in this study where in general both lattice parameter and B increase with increase in Nb, despite the surprising decrease in B at 6.25 at.% Nb. This observation can be explained using valence electron density theory. Previous studies have reported that B has a direct relationship with valence electron density, that is, more electrons will correspond to greater repulsions within a structure [41]. The bulk moduli for the

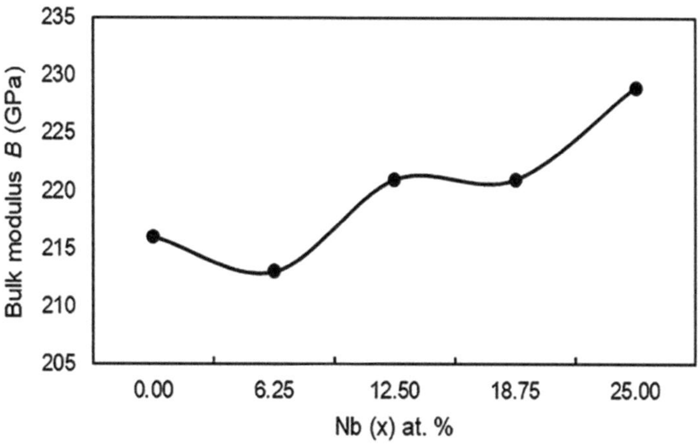

Figure 6.
Bulk modulus B *of B2* $Ti_{50-x}Nb_xRu_{50}$ *alloys.*

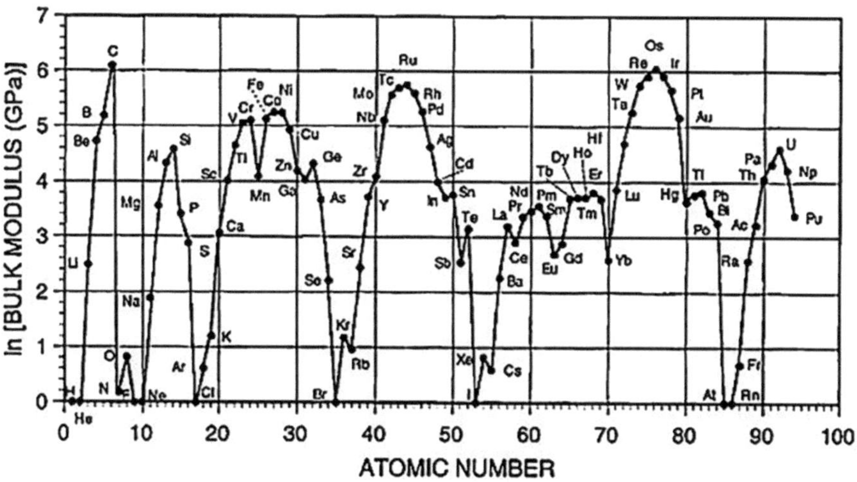

Figure 7.
Natural logarithm of bulk modulus as a function of atomic number [42].

first 94 periodic table elements are shown in **Figure** 7 [43]. It is shown that the *B* of Ru is amongst the highest values and that of Nb is greater than *B* of Ti. Also, it can be seen that there are large variations in moduli with atomic number. The maxima is observed when the valence shells of rows one and two are half full electrons and when the transition shells d-orbitals (3d, 4d 5d) are roughly half filled.

3.3 Shear modulus, *G*

Shear modulus is a measure of resistance against changes in atomic bond length and angle or resistance to shape change. It can be expressed using Eq. (6) [40]. A larger shear modulus means greater ability to resist shearing forces. **Figure 8** shows that shear modulus decreases with increasing Nb content. At first, Nb additions below 12.50 at.% show small change less than 5 GPa, then a significant decrease of about 15 GPa is observed as the Nb content is increased beyond that. This demonstrates that $Ti_{50}Ru_{50}$ has a higher tendency to resist motion of planes within a solid parallel to each other [42]. This is the reason why such an alloy is not desirable for development of superelastic materials because from crystallographic point of view, the transformation from B2 to low temperature phase occurs by Bain strain and lattice-invariant shear. The Bain strain consist of atomic movements needed to produce new structure from the parent phase, while the lattice invariant shear is an accommodation step.

The general mechanisms by which this happens is either by slip or twinning. Twinning is unable to accommodate volume changes but, accommodation is achieved by shape change in a reversible way. Therefore, here we observe that the addition of Nb is necessary in order to induce (to a certain degree) shear deformation that is required for phase transformation to occur by twinning.

$$G = \frac{1}{2}\left[\frac{C_{11} - C_{12} + 3C_{44}}{5} + \frac{5C_{44}(C_{11} - C_{12})}{4C_{44} + 3(C_{11} - C_{12})}\right] \tag{6}$$

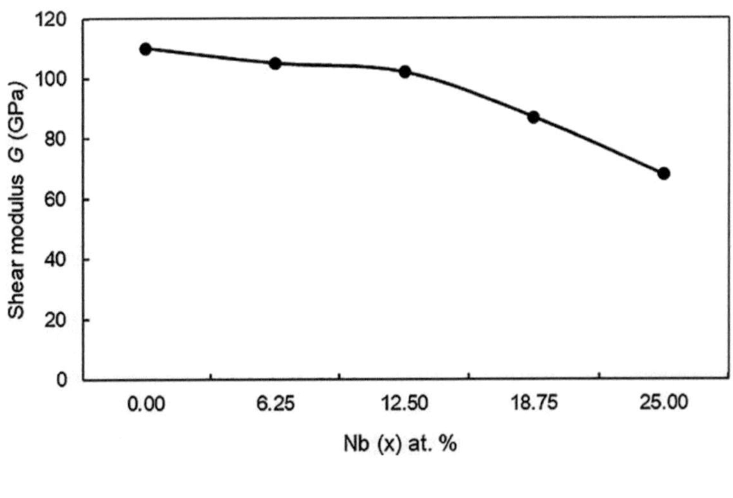

Figure 8.
Shear modulus G of B2 $Ti_{50-x}Nb_xRu_{50}$ alloys.

3.4 Young's modulus, *E*

Young's modulus (*E*) is the measure of a material's ability to withstand changes in length when under tension or compression (material's stiffness). This mechanical feature is important in biomedical materials for determining the specific implant application. For instance, if an implant has a greater *E* than the replaced bone, stress shielding occurs. This stress shielding effect prohibit the transfer of needed stress to the adjacent bone, which leads to bone resorption around the implant and consequently cause death of bone cells. Therefore, an ideal biomedical material should have combined properties of modulus closer to that of bone and high strength to sustain a long term service period and reduce revision surgery. **Figure 9** shows a typical compressive *E* of different types of human soft bone [44].

Apart from transformation strain, it is also well established that SE is dependent on the *E* of the material. The degree of SE is typically small for alloys with high *E* compared to ones having low *E*, as shown in a schematic diagram in **Figure 10** (Spring back = SE) [45]. Here, the theoretical elastic constants data have been used to predict

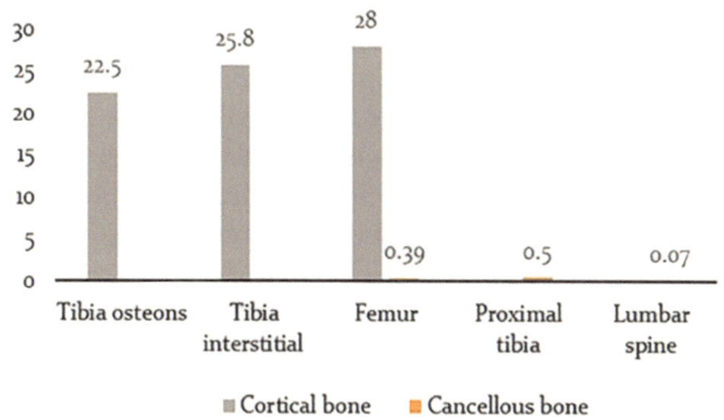

Figure 9.
Young's modulus E *(GPa) of human bone [44].*

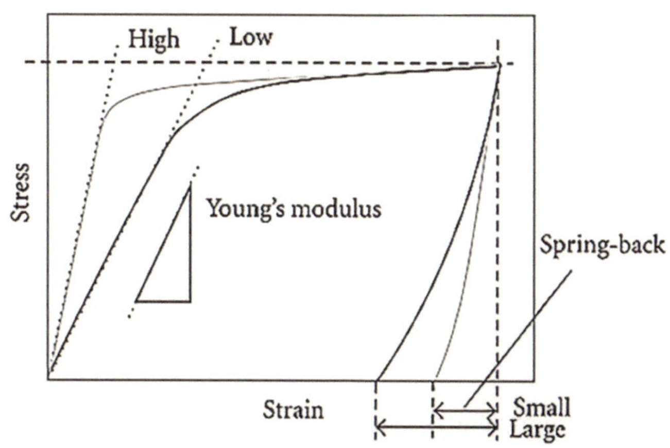

Figure 10.
The relationship between E *and SE [45].*

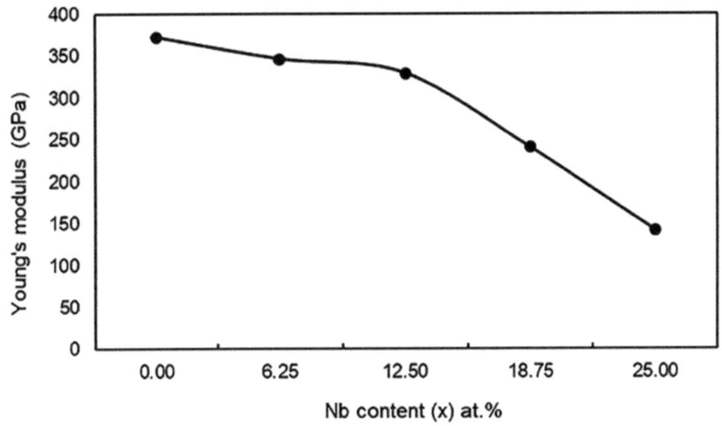

Figure 11.
Young's modulus E of B2 $Ti_{50-x}Nb_xRu_{50}$ alloys.

the *E* of B2 $Ti_{50-x}Nb_xRu_{50}$ at 0 K. **Figure 11** shows the *E* across different compositions. The composition dependence trend is similar to that of shear modulus, with highest value being 372 GPa for $Ti_{50}Ru_{50}$. Again, further substitution of Ti with Nb reduces the *E* to less than 150 GPa. Based on the predicted trend, it is anticipated that increasing Nb composition further than considered in this study might result in *E* closer to 50 GPa which will be closer to that of B2 TiNi and human bone [46, 47]. As a result, the current predicted data is encouraging to pursue this work further.

On contrary to what has been believed to be a disadvantage in terms of high *E*, recently other researchers have found that high *E* can be of advantage if all other biomedical aspect are taken into consideration. For example, Nakai et al. [45] discussed that surgeons specializing in spinal diseases pointed out that the amount of SE in implant rods must be balanced such that it offers better handling during operation, but also be ductile enough not to create stress shielding for the patient. Moreover, biomaterials with higher *E* and consequent high yield strength have received much attention lately in the development of alternative porous orthopedic implants via powder metallurgy routes such as additive manufacturing (AM) techniques [48]. The porosity route is viewed as one of the viable means to reduce *E* and thus overcome stress-shielding health-risk.

Therefore, in order to satisfy both surgeon and patient's requirements, the *E* should be adjusted by phase transformations. That is, during deformation, phase transformation occurs such that the new phase introduces high *E*, while the non-deformed part remains low. That is, in orthopedic operations for treating spinal diseases, the implant rod is bent so that it corresponds to the curvature of the spine. Thus, an alloy with desired superelastic properties can be designed to suit the surgeon's requirement. Then during operation, while the material is bent, the new deformation-induced phase with high Young's modulus is formed for surgeon's requirement and the non-deformed part remains low for patient comfort. The illustration in **Figure 12** demonstrates how the Young's modulus can be adjusted in β alloys [45]. This can similarly be true for B2 alloys, as we have observed that the effect of Ru is the same in both β and B2 alloys, although with B2 alloys, the deformed part may need to have lower *E*.

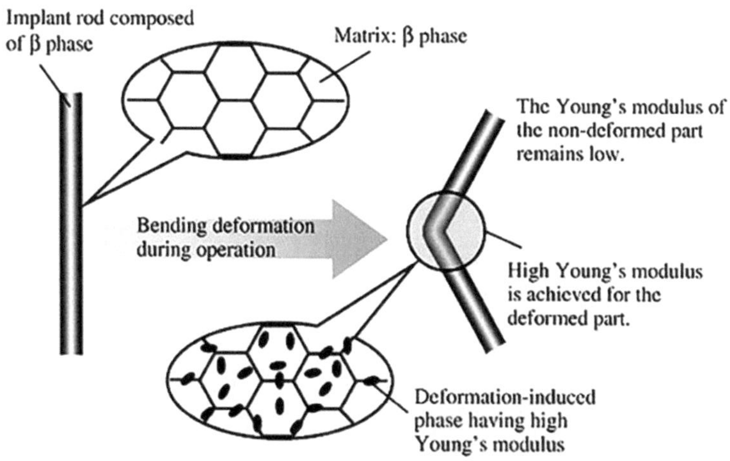

Figure 12.
Illustration of self-adjustment of Young's modulus in implant rod [45].

3.5 Stiffness ratios

The brittle/ductile behavior of materials can be determined using the Pugh's modulus ratio $k = G/B$ [40]. The k ratio highlights the relationship between the elastic and plastic properties of crystalline materials, where brittle materials have high k values (>0.57) and ductile materials have low ones (<0.57) [41]. It is known that with approximately equal values of shear G and bulk B moduli under applied tensile stress, atomic bonds in local volume tend to dilate and rotate. This occurs instead of the domination of uniform rotation that leads to shear deformation in the plastic stage in crystals whose B is significantly greater than G. This means that atoms in local volume undergo a random movement under applied stress and if strain rate exceeds the rate of atomic relaxation, the amorphous state may result [49].

Moreover, in simple tension or compression tests, a force is applied to a rod parallel to its axis creating a tensile or compression stress. The rod responds by either elongating (tensile) or shortening (compression), leading to change in axial strain ($\pm\varepsilon = \Delta l/l$) to a fractional amount. This results in simultaneous decrease (or increase under compression) of its cross-sectional area. The ratio of transverse contraction, $-(\Delta d/d)$, to the longitudinal extension, $(+\varepsilon = \Delta l/l)$, is the Poisson ratio υ [43].

In order to predict the stiffness ratio of B2 $Ti_{50-x}Nb_xRu_{50}$ alloy composition, we used both Pugh's ratio and Poisson ratio. **Figure 13** shows the calculated ratios with respect to composition. As expected, the k ratio plot trend shows that B2 $Ti_{50}Ru_{50}$ is more brittle than presently considered ternary alloy compositions, as indicated by decreasing values. This agrees well with experimental data that was reported by Tsuji et al. [3], where TiRu was measured to have hardness of about 400 GPa and it was deduced that the addition of Ru to TiNi increased hardness and thus, decreasing ductility. Thus we see that high Ru alloys are brittle, leading to the need to introduce Nb at the expense of reducing Ti in order to induce ductility. The Poisson ratio increases with increasing Nb content, further indicating the improved ductility with Nb additions.

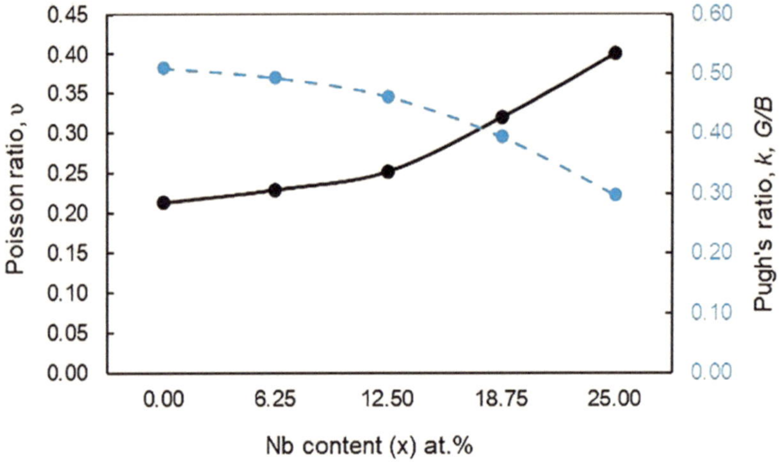

Figure 13.
The calculated stiffness ratios of B2 $Ti_{50-x}Nb_xRu_{50}$ *alloys.*

Device type	Ru-effect on properties
Spinal fixation devices	• Increased Young's modulus reduces the spring-back during the operation [45, 50] • Ultra-low magnetic susceptibility [2]
Removable internal fixation devices, e.g. bone plates	• Inhibition of precipitation of calcium phosphate, thus allowing easy removal of an implant at the end of healing process [51]
Orthopedic implants	• Addition of Ru in βTiNb alloys increased tensile and fatigue strength [2, 4] • Stability of high temperature phase (β or B2) to low temperatures improves possibility of superelasticity [10, 19, 21, 52]
Vascular stents	• Ultra-low magnetic susceptibility [2] • Improved corrosion resistance [21]

Table 3.
Possible applications of Ru-rich biomaterials.

3.6 Possible biomedical applications of Ru-rich alloys

The possible applications of Ru-rich alloys in biomedical filed are summarized in **Table 3**. The investigated alloys showed that the Young's modulus of about 100 GPa is attainable, which is attributed to high Ru content. The stiffness ratios also indicated that even with high Ru content the alloy can still be mechanically adjusted to fit in the application criteria. For example, an alloy composition with such high Young's modulus can be beneficial during the spinal fixation operations. Previously, surgeons had to rely on the phase transformation of β-phase to brittle ω-phase precipitates to prevent the occurrence of superelasticity within the limited time of adapting the spinal fixation device to the physiological curvature of the spine [45, 51]. The currently investigated alloys are structurally more ordered due to stabilizing effect of Ru, and thus the phase transformation to brittle ω-phase is suppressed. The result is that the Young's modulus does not drastically change during deformation. It is only by continuous cyclic loading after the installation of the implant that the transformation of B2 phase to martensite phase can be observed and this is where the patient can receive the benefit of precipitated soft phase to accommodate the surrounding human tissue and prevent stress-shielding effect. Therefore, this means that the correct alloy composition of B2 $Ti_{50-x}Nb_xRu_{50}$ can be developed for use in the design of spinal fixation devices.

Other possible applications of Ru-rich biomaterials is in removable implants such as the internal fixation devices implanted into the bone marrow (e.g. femoral, tibia and humeral marrow), screws used for bone plate fixation and implants used for children. These type of implants may grow into the bone, but it is essential to remove the internal fixation device after surgery, owing to specific local symptoms indications such as palpable hardware, wound exposure of hardware and also the implant may need to be removed in the case where an athlete return to contact sports. Surgeons have experienced difficulties with the removal of these fixation devices because during the assimilation of the device with the bone there is precipitation of calcium phosphate which might cause bone refracture [53].

Thus, in this instance, it is important to prevent the adhesion of the alloys with the bone tissues. This implies that the newly developed alloy must be able to inhibit the precipitation of calcium phosphate. The reactivity of Ru-containing alloys have been found to decrease with increasing Ru-content in tetrazolium salt (MTS) assay, which represents the body fluid [2, 4]. Also Ti has been observed to quickly facilitate the formation of calcium phosphate in order to improve bone adherence. Previously, one had to coat the Ti substrate with the Zr coating which could have resulted in other properties being compromised. Thus for the current alloys with high Ru content, the precipitation of calcium phosphate can be prevented because the Ti content has been reduced and the Ru is non-toxic, allergy-free and has the potency to prevent quick reaction of Ti and consequently reduce deposition of calcium phosphate [53].

Furthermore, in cardiology, the imaging technology can begin with the advent of chest X-ray, in which a topographic image of a heart could be done in different projections. This is often done by multiple projections to help analyze cardiac structures and the location of abnormalities. It has been reported that most conventional metallic biomaterials with high magnetic susceptibility have adverse effect on the MRI and diagnostics. The resulting artifacts can distort the authentic bio-imaging of the human organs and tissues around the implant. Materials and devices with an ultra-low magnetic susceptibility are required for surgery and diagnostics performed under MRI [2]. Addition of Ru can improve/lower magnetic susceptibility as observed in the Zr-Ru alloys where an ultra-low magnetic susceptibility was obtained with increasing Ru content [2].

4. Conclusions

In this study the B2 TiRu alloy has been modified in order to improve the underlying mechanical properties required in biomaterials. The advantages of Ru addition have been outlined, which include biocompatibility, improved corrosion resistance, β-stabilizer and also B2 stabilizer and ultralow magnetic susceptibility. However, high Ru content of about 50 at.% results in a more ordered structure, which has adverse effects on the mechanical properties. Therefore, although it may be beneficial to have a high Ru alloy, it is also crucial to identify the shortfalls arising due to Ru addition.

The approach used started with TiRu binary compound, which is known to be hard and brittle. In order to induce beneficial mechanical properties for biomedical applications, the systematic substitution of Ti with Nb was performed.

The results showed that 50 at.% Ru constitute to high stability of B2 phase at 0 K, such that even with increasing Nb content, B2 phase remained highly stable (C' positive). The bulk modulus showed variation with increasing Nb content, whilst the shear modulus decreased with increasing Nb content. Also, of importance to the design of biomaterials is SE, which is dependent on both the transformation strain and Young's modulus. The Young's modulus decreased with increasing Nb content. Therefore, in order to design an alloy with beneficial properties for both surgeons and patients, we propose higher Nb content towards potential high temperature shape memory alloy B2 $Nb_{50}Ru_{50}$. Current results are encouraging that by increasing Nb content it could be possible to identify ternary B2 $Ti_{50-x}Nb_xRu_{50}$ composition that has Young's modulus closer to 50 GPa which will be closer to that of B2 TiNi and human bone. This means that we are able to retain high Ru with said biomedical properties, whilst achieving the desired mechanical properties.

Acknowledgements

The authors acknowledge the Advanced Materials Initiative (AMI) of the Department of Science and Innovation (DSI) through Mintek for their financial support. The authors are also grateful to the National Research Foundation (NRF) South Africa—JSPS GRANT No: 148782. The gratitude is also extended to the Centre for High-Performance Computing (CHPC) in Cape Town for allowing us to carry out the calculations using their remote computing resources.

Conflict of interest

The authors declare no conflict of interest.

Author details

Duduzile Nkomo[1,2*], Maje Phasha[1] and Hein Moller[1,2]

1 Mintek, Advanced Materials Division, Johannesburg, Randburg, South Africa

2 Department of Materials Science and Metallurgical Engineering, University of Pretoria, Pretoria, Hatfield, South Africa

*Address all correspondence to: dudunk@mintek.co.za

References

[1] Murray J. The Ru-Ti (ruthenium-titanium) system. Bulletin of Alloy Phase Diagrams. 1982;**3**(2):216-221

[2] Li HF, Zhou FY, Li L, Zheng YF. Design and development of novel MRI compatible zirconium-ruthenium alloys with ultralow magnetic susceptibility. Scientific Reports. 2016;**6**(April):1-10

[3] Masahiro T, Hideki H, Wakashima K, Yamabe-Mitarai Y. Phase stability and mechanical properties of Ti(Ni, Ru) alloys. Mat. Res. Soc. Symp. Proc. 2003; **754**:1-6

[4] Biesiekierski A, Lin J, Li Y, Ping D, Yamabe-Mitarai Y, Wen C. Impact of ruthenium on mechanical properties, biological response and thermal processing of β-type Ti–Nb–Ru alloys. Acta Biomaterialia. 2017;**48**:461-467

[5] Chen BS, Liu JL, Wang C, Guan XY, Song JQ, Li YZ. Phase stability, elastic properties and martensitic transformation temperature of Zr50Pd50-xRux alloys from first-principles calculations. Journal of Alloys and Compounds. 2016;**662**:484-488

[6] Otsuka K, Ren X. Physical metallurgy of Ti-Ni-based shape memory alloys. Progress in Materials Science. 2005; **50**(5):511-678

[7] Kim HY, Ikehara Y, Kim JI, Hosoda H, Miyazaki S. Martensitic transformation, shape memory effect and superelasticity of Ti-Nb binary alloys. Acta Materialia. 2006;**54**(9): 2419-2429

[8] Kim HY, Miyazaki S. Martensitic transformation and superelastic properties of Ti-Nb base alloys. Materials Transactions. 2015;**56**(5): 625-634

[9] Al-Zain Y, Kim HY, Hosoda H, Nam TH, Miyazaki S. Shape memory properties of Ti-Nb-Mo biomedical alloys. Acta Materialia. 2010;**58**(12): 4212-4223

[10] Al-Zain Y, Sato Y, Kim HY, Hosoda H, Nam TH, Miyazaki S. Room temperature aging behavior of Ti-Nb-Mo-based superelastic alloys. Acta Materialia. 2012;**60**(5):2437-2447

[11] Bönisch M et al. Thermal stability and phase transformations of martensitic Ti-Nb alloys. Science and Technology of Advanced Materials. 2013;**14**(5):055004

[12] Hao YL, Li SJ, Sun SY, Zheng CY, Yang R. Elastic deformation behaviour of Ti-24Nb-4Zr-7.9Sn for biomedical applications. Acta Biomaterialia. 2007; **3**(2):277-286

[13] Karre R, Niranjan MK, Dey SR. First principles theoretical investigations of low Young's modulus beta Ti-Nb and Ti-Nb-Zr alloys compositions for biomedical applications. Materials Science and Engineering: C. 2015;**50**:52-58

[14] Kumar H, Rajamallu K, Tamboli RR, Dey SR. Fabrication of beta Ti29Nb13Ta4.6Zr alloy through powder metallurgy route for biomedical applications. Powder Metallurgy. 2020; **2018**

[15] Ramarolahy A, Castany P, Prima F, Laheurte P, Péron I, Gloriant T. Microstructure and mechanical behavior of superelastic Ti-24Nb-0.5O and Ti-24Nb-0.5N biomedical alloys. Journal of the Mechanical Behavior of Biomedical Materials. 2012;**9**:83-90

[16] Zhang DC, Lin JG, Jiang WJ, Ma M, Peng ZG. Shape memory and superelastic behavior of Ti-7.5Nb-4Mo-

1Sn alloy. Materials and Design. 2011;**32** (8–9):4614-4617

[17] Zhang J et al. Influence of equiatomic Zr/Nb substitution on superelastic behavior of Ti-Nb-Zr alloy. Materials Science and Engineering A. 2013;**563**: 78-85

[18] Kim HS, Kim WY, Lim SH. Microstructure and elastic modulus of Ti-Nb-Si ternary alloys for biomedical applications. Scripta Materialia. 2006; **54**(5):887-891

[19] Biesiekierski A, Ping DH, Yamabe-Mitarai Y, Wen C. Impact of ruthenium on microstructure and corrosion behavior of β-type Ti-Nb-Ru alloys for biomedical applications. Materials and Design. 2014;**59**:303-309

[20] Mohd Jani J, Leary M, Subic A, Gibson MA. "A review of shape memory alloy research, applications and opportunities". Materials and Design. 2018;**56**(January):1078-1113. 2014

[21] Biesiekierski A, Wang J, Abdel-Hady Gepreel M, Wen C. A new look at biomedical Ti-based shape memory alloys. Acta Biomaterialia. 2012;**8**(5):1661-1669

[22] Wadood A, Takahashi M, Takahashi S, Hosoda H, Yamabe-Mitarai Y. High-temperature mechanical and shape memory properties of TiPt-Zr and TiPt-Ru alloys. Materials Science and Engineering A. 2013;**564**:34-41

[23] Waterstrat RM, Kuentzler R. Structural instability in the B2-type ordered alloys Zr (Ru,Rh) and Zr (Ru, Pd). Journal of Alloys and Compounds. 2003;**359**(1–2):133-138

[24] Manzoni AM, Denquin A, Vermaut P, Puente Orench I, Prima F, Portier RA. Shape memory deformation mechanisms of Ru-Nb and Ru-Ta shape memory alloys with transformation temperatures. Intermetallics. 2014;**52**:57-63

[25] Notomi M, Van Vliet KJ, Yip S. Classification and characterization of the shape memory binary alloys. Materials Research Society Symposium Proceedings. 2007;**980**:223-228

[26] Ngobe B, Phasha M, Mwamba I. First-principles study to explore the possibility of inducing martensitic transformation in ordered B2 TiRu phase by alloying with Pd. Suid-Afrikaanse Tydskr. vir Natuurwetenskap en Tegnol. 2022;**40**(1):205-211

[27] Bai X, Li JH, Dai Y, Liu BX. Structural and Elastic Properties of Pd-Zr Compounds Studied by Ab Initio Calculation. Journal Intermet. 2012

[28] Hu JQ, Xie M, Pan Y, Yang YC, Liu MM, Zhang JM. The electronic, elastic and structural properties of Pd-Zr intermetallic. Computational Materials Science. 2012;**51**(1):1-6

[29] Clark SJ et al. First principles methods using CASTEP. Zeitschrift fur Krist. 2005;**220**(5–6):567-570

[30] Vanderbilt D. Soft self-consistent pseudopotentials in a generalized eigenvalue formalism. Physical Review B. 1990;**41**:7892-7895

[31] Perdew JP, Wang Y. Accurate and simple analytic representation of the electron-gas correlation energy. Physical Review Letters. 1992;**77**:3865-3868

[32] Perdew JP, Burke K, Ernzerhof M. Generalized gradient approximation made simple. Physical Review Letters. 1997;**77**(18):3865-3868

[33] Bönisch M et al. Giant thermal expansion and α-precipitation pathways

in Ti-alloys. Nature Communications. 2017;**8**(1):1-9

[34] "Periodic table of elements" [Online]. Available from: https://www.vertex42.com/ExcelTemplates/periodic-table-of-elements.html. [Accessed: January 2023]

[35] Wilson JA et al. Predicting the thermal expansion of body-centred cubic (BCC) high entropy alloys in the Mo-Nb-Ta-Ti-W system. Journal of Physics: Energy. 2022;**4**(3):034002

[36] Mouhat F, Coudert FX. Necessary and sufficient elastic stability conditions in various crystal systems. Physical Review B: Condensed Matter and Materials Physics. 2014;**90**(22):0-3

[37] Mehl MJ, Singh DJ, Papaconstantopoulos DA. Properties of ordered intermetallic alloys: First-principles and approximate methods. Materials Science and Engineering A. 1993;**170**(1–2):49-57

[38] Mahlangu R, Phasha MJ, Chauke HR, Ngoepe PE. Structural, elastic and electronic properties of equiatomic PtTi as potential high-temperature shape memory alloy. Intermetallics. 2013;**33**: 27-32

[39] Li X. Doc-Mechanical Properties of Transition Metal Alloys from First-Principlestheory. Doctoral dissertation, KTH Royal Institute of technology. 2015

[40] Phasha MJ, Ngoepe PE, Chauke HR, Pettifor DG, Nguyen-Mann D. Link between structural and mechanical stability of fee- and bcc-based ordered Mg-Li alloys. Intermetallics. 2010; **18**(11):2083-2089

[41] Tian Y, Xu B, Zhao Z. Microscopic theory of hardness and design of novel superhard crystals. International Journal of Refractory Metals and Hard Materials. 2012;**33**:93-106

[42] Xie M. High Pressure Studies of Ultra-Incompressible, Superhard Metal Borides. Los Angeles: University of California; 2013

[43] Gilman JJ, Cumberland RW, Kaner RB. Design of hard crystals. International Journal of Refractory Metals and Hard Materials. 2006;**24** (1–2):1-5

[44] Zhang LC, Chen LY. A review on biomedical titanium alloys: Recent progress and prospect. Advanced Engineering Materials. 2019;**21**(4):1-29

[45] Nakai M, Niinomi M, Zhao X, Zhao X. Self-adjustment of Young's modulus in biomedical titanium alloys during orthopaedic operation. Materials Letters. 2011;**65**(4):688-690

[46] Filip P. Titanium-Nickel Shape Memory Alloys in Medical Applications. 2001. pp. 53-86

[47] Agnesh Rao K, Kothari A, Prakash DL, Hallikeri JS, Shetty NV. A review on nickel titanium and it 's biomedical applications. International Research Journal of Engineering and Technology (IRJET). 2020:1742-1748

[48] Limmahakhun S, Oloyede A, Sitthiseripratip K, Xiao Y, Yan C. Stiffness and strength tailoring of cobalt chromium graded cellular structures for stress-shielding reduction. Materials and Design. 2017;**114**:633-641

[49] Yang R, Hao Y, Li S. Development and application of low-modulus biomedical titanium alloy Ti2448. Biomedical Engineering Trends in Materials Science. 2011;**49**(10): 225-247

[50] Niinomi M. Recent progress in research and development of metallic structural biomaterials with mainly focusing on mechanical biocompatibility. Materials Transactions. 2018;**59**(1):1-13

[51] Niinomi M, Nakai M. Titanium-based biomaterials for preventing stress shielding between implant devices and bone. International Journal of Biomaterials. 2011;**2011**:1-10

[52] Eisenbarth E, Velten D, Müller M, Thull R, Breme J. Biocompatibility of β-stabilizing elements of titanium alloys. Biomaterials. 2004;**25**(26):5705-5713

[53] Kobayashi E et al. Inhibition effect of zirconium coating on calcium phosphate precipitation of titanium to avoid assimilation with bone. Materials Transactions. 2007;**48**(3):301-306

Section 3

Advanced Applications

Chapter 5

Electrochemical Investigation of Heat Treated PtRu Nanoparticles Prepared by Modified Polyol Method for Direct Methanol Fuel Cell Application

Adebare Nurudeen Adewunmi, Ntalane Sello Seroka, Su Huaneng and Khotseng Lindiwe Eudora

Abstract

In this work, heat-treated PtRu metal alloys based on multi-walled carbon nanotubes (MWCNT) were synthesized using modified polyol approach for methanol oxidation reaction (MOR) in acidic conditions at 2500, 3500, and 4500 C. The catalysts physical and electrochemical properties were investigated. The High Resolution Transmission Electron Microscopy (HR-TEM) was used to determine the shape, particle size, and particle size distribution of the catalysts, where spherical and agglomerated PtRu nanoparticles with narrow particle size distribution were observed with particle sizes ranging from 0.600 to 1.005 nm. Their crystalline sizes were assessed using the XRD with catalysts presenting a face-centered crystal structure, which is typical of platinum structures with crystalline sizes ranging from 0.500 to 1.180 nm. Energy-Dispersive Spectroscopy, (EDS), was used to identify the elements. Cyclic voltammetry (CV) was used to determine the electroactive surface area (ECSA) and MOR of the electrocatalysts, whereas electrochemical impedance spectrometry (EIS) and chronoamperometry (CA) were used to study their electro-kinetics and stability towards MOR, respectively. PtRu/MWCNT electrocatalysts alloyed at 450°C showed better electroactivity and kinetics as compared to other catalysts, evident from the highest current density of 19.872 mA/cm^2 and lowest charge transfer resistance of 0.151 kΩ from CA and EIS, respectively.

Keywords: methanol oxidation, catalysts, multi-walled carbon nanotubes, heat treatment, electroactivity

1. Introduction

A device called a fuel cell produces electricity through chemical processes. The cell can continue to generate power as long as fuel is available [1, 2]. In order to provide sustainable electricity, fuel cells combine well with other clean, contemporary energy sources including wind, solar, and hydroelectricity. These devices have a variety of uses in portable devices, stationary equipment, and transportation, as well as silent operation without vibration and inherent modularity that enables simple design [3].

The versatile and mutual conversion of chemical to electrical energy via fuel cell technologies has been considered a green approach and environmentally friendly than the combustion of fossil fuels. The combustion of fossil fuel to generate energy has contributed to acid rain, ozone depletion and climate change. Furthermore, this is as a result of air pollution and harmful greenhouse gases such as CO2 [2, 4]. The Kyoto Protocol has since then suggested the use of renewable energy sources, the promotion of existing high efficiency electricity technologies, and the adoption of advanced low-CO2 emission energy systems [2, 5], there has been intensive research and development of renewable and environmentally friendly ways to generate electricity.

Electrochemical systems, such as fuel cells, effectively transform chemical energy directly into electricity, with water and heat as byproducts [3]. The device is composed of four major components: anode and cathode electrodes, an electrolyte, and a gas diffusion layer. The anode electrode of the device receives fuel, whereas the cathode electrode receives oxygen or air. The basic objective of an electrolyte located between the electrodes, regardless of the type of cell, is always to move ions (anions or cations) from one side to the other [6].

In the early 1800s, Sir William Grove reported the first fuel cell when he built the gas battery, a device that combined hydrogen and oxygen to produce electricity. This technology was later referred to as a fuel cell. Francis Thomas Bacon demonstrated the first fully working fuel cell in 1959 [3, 4]. There are various varieties of fuel cells accessible today, each having its unique chemical fuel input and operating principles (**Figure 1**).

For a hydrogen fuel cell, similar cathode and anode reactions are

Cathode: $1/2O_2 + 2H^+ + 2e- \rightarrow H_2O$.

Anode: $H_2 \rightarrow 2H^+ + 2e-$.

Overall reaction: $1/2O_2 + H_2 \rightarrow H_2O$.

1.1 Fuel cells technologies

The choice of an electrolyte used in fuel cells is the most significant factor and gives it an identity. The operating conditions such as temperature range, catalyst desired, type of fuel, and other parameters have an effect on the electrochemical processes taking place in the cell due to the distinction of electrolytes [8]. Thus qualities influence the applications for which these cells are best suited. Various fuel cell types are now being explored, each with their own set of advantages, limitations, and prospective applications [8]. PEMFCs, direct methanol fuel cells, alkaline fuel cells, phosphoric acid fuel cells, solid oxide fuel cells, and molten carbonate fuel cells are all examples of fuel cells.

1.2 Catalyst

A catalyst helps to speed up a chemical reaction by creating bonds with the molecules involved. The catalyst permits the reactants to react to produce a product, which then detaches from the catalyst and leaves it undamaged for the next reaction. As a result, a catalytic reaction may be thought of as a cyclic event in which the catalyst precipitates and then returns to its original state at the conclusion of the cycle.

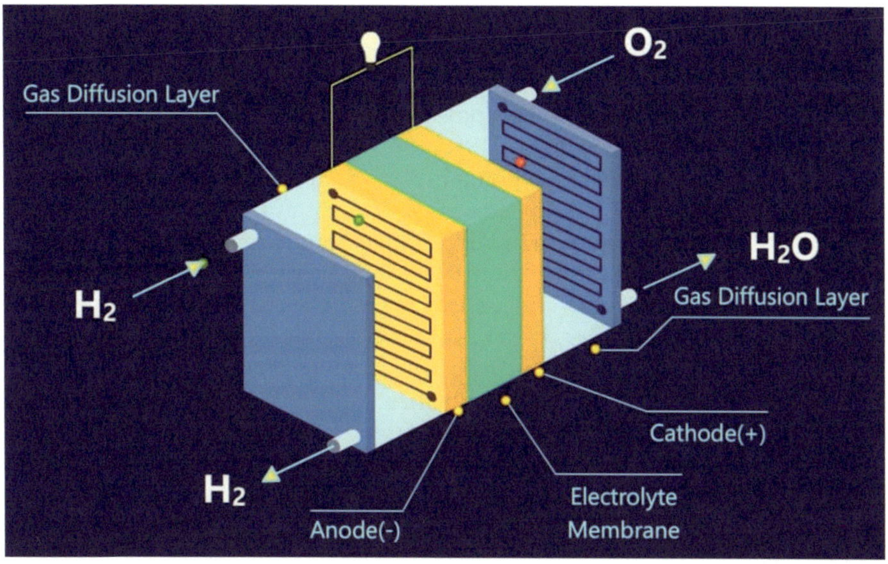

Figure 1.
A schematic diagram showing a hydrogen based fuel cell [7].

Catalysts are the workhorses of chemical transformations in industry, accounting for 85–90 percent of total chemical output. A catalyst provides an alternative, more energy-efficient mechanism to the non-catalytic reaction, allowing activities to be carried out at pressures and temperatures that are industrially practical.

1.3 Fuel cell electrocatalysts

Catalysts are used in fuel cells to increase reaction kinetics in the anode and cathode sectors. Efficient electrocatalysts are needed in fuel cells to increase cell's performance, notably its durability, stability, and activity, as well as to lower cost. PtRu catalysts are used due to its high CO tolerance through bifunctional mechanism. Because the catalysts used in fuel cells are noble metals and non-noble immobilized on a conducting support, the catalysts utilized in fuel cells are classed as heterogeneous. The reactants and the heterogeneous catalyst are in distinct phases.

1.4 Methanol oxidation anode catalysts

Pt, Pd, and Au are well-known precious metals that can be used as methanol oxidation catalysts in DMFCs [9–12]. As a result, these precious metal catalysts have received a lot of attention in DMFCs [9–11]. Saving money and increasing methanol oxidation activity are both appealing characteristics. The morphological variations in particle size distribution, shape, and surface structure of precious metal catalysts have a strong influence on their methanol oxidation activity [10]. Many teams have recently created better morphologies of noble metal catalysts to promote methanol oxidation activity with low metal loading [10, 11]. Platinum-based catalysts have been shown to be reactive and stable in the acidic DMFC environment [12, 13].

Another significant barrier to efficient methanol fuel-to-electric-current conversion in a DMFC is the anode catalyst's poor MOR kinetics. Surface poisoning by chemical intermediates such as COads-like species formed during the stepwise dehydrogenation of methanol is primarily responsible for this delay [14–16].

The importance of DMFC's low efficiency and overpotential stems from oxidized adsorbed COads-like species and other poisoning intermediates for probable MOR on Pt at such potentials [17]. PtRu [18, 19], PtSn [20, 21], PdNi [22], PtMo [23, 24], PtTiO2 [25, 26], PtW [23], PtOs [27], and PtMn [28] are examples of binary Pt-based alloy (double-component) catalysts. The concept of metal alloys is to enhance methanol electro-oxidation activity, by supplying a flux of hydroxyl species to negative electrode at higher potentials to facilitate the reaction with COads-like poisoning species. The methanol oxidation and associated mechanisms have been studied under controlled conditions in a variety of catalytic systems [29–37].

Platinum's electroactivity appears to be influenced by its morphology [38], with roughened platinum having significantly higher activity [39].

Several researchers have looked into the effect of particle size on methanol oxidation [40–42]. For particles with a diameter of less than 5 nm, several scientists have found a decline in activity as the particle size decreases [43–46].

On a pristine platinum electrode, linearly bound CO can cover up to 90% of the active sites, blocking the majority of them. These findings have been frequently confirmed by various researchers [47–49].

Platinum has been looked at a lot as an electrocatalyst for the oxidation of methanol. To remain being active it is used as a composite to avoid formation of CO. This has led to the search for other active materials, especially those that could work with platinum as a promoter by making it easier for chemisorbed CO to be oxidized [13].

Ru has been discovered to enhance anti-poisoning effectiveness. The Pt-Ru catalyst's high price, however, is one of the main obstacles to its broad implementation [50]. Reduced catalyst costs are a key factor in the widespread commercialization of DMFCs [51]. A number of organizations have worked diligently to create new multi-component catalyst systems for the oxidation of methanol, including PtRuNi [52–54], Pt–Ru–Os [55, 56], Pt–Ru–Mo–W [56] and Pt–Ru–Sn–W [57]. Investigations on the Pt-Ru-Sn system were also conducted [58], but it was found that adding tin to the Pt-Ru alloy causes the Ru to be evacuated with minimal advantage [59].

Fuel cells have received a lot of attention among other energy technologies due to their high rates of energy conversion, cheap availability to fuel, and environmental friendliness [60, 61].

Fuel cells use redox processes involving oxygen and fuels to turn chemical energy into electrical energy [62, 63]. Despite years of hard work, there are still certain fundamental concerns that must be addressed. Platinum is now the most popular MOR catalyst, and significant efforts have been undertaken to increase its activity and utilization. One of the numerous disadvantages of pt-based catalysts is that they are sensitive to impurities and pricey. The main methodology used in this work was alloying platinum with ruthenium metal via the modified polyol method of catalyst synthesis, and the resultant nanoparticles were subsequently heated at higher temperatures to improve the methanol electrooxidation reaction [64–66].

Because heat treatment has been shown to affect the activity of electrocatalysts, it is known as thermal activation. Heat treatment of Pt-based catalysts can result in particle-size increase, improved alloying, and changes in the surface morphology of the catalyst from amorphous nature to more crystallite phases, all of which have a significant impact on their electroactivity and stability [67].

Numerous studies [67] have been conducted to determine how heat treatment affects Pt catalysts. In this study, heat treatments of 2500C, 3500C, and 4500C are applied to binary PtRu/MWCNT catalysts made using the modified polyol method to see how they affect their shape, particle size, electroactivity, and stability towards methanol electrooxidation.

According to Valisi et al. [68], enhancing the electrocatalytic activity and stability of fuel cells, thermal treatment of catalysts plays an integral part of ORR. The ORR activity was found to be maximal as indicated, Pt/C (T, 350°C) < Pt-Co/C (T, 250°C) < Pt-Ni/C (T, 350°C) < Pt-Fe/C (T, 350°C) < Pt-Cu/C (T, 350°C).

Xiowel et al. [69] produced carbon-supported Au-PtRu catalysts for DMFC by easily depositing Au metal on top of a commercial PtRu/C catalyst and then heating it at three different temperatures. After a straightforward Au particle deposition on a commercial Pt-Ru/C catalyst, the resulting composite catalyst was heated at 125, 175 and 200 degrees Celsius in a N environment. It was discovered that among heat-treated catalysts, Au-PtRu/C catalysts had the greatest electrocatalytic stability.

Another study [70] found that heat treatment enhanced the electrical conductivity of the electrode, which influenced the agglomeration of Pt particles. Furthermore, the PtM/C alloying degree (M denotes Cr, Pd, Co) was greatly enhanced. Finally, the PEM fuel cell displayed an appealing performance, with PtCo/C delivering an approximate current density of 392.8 mA/Cm^2.

The elimination of any undesired contaminants that may have resulted from the early phases of preparation is one of the benefits of heat treatment. Heat treatment increases the electrocatalytic activity of the produced catalyst by allowing for uniform dispersion and stable metal distribution on the substrate [71, 72]. Metal particle size and distribution, particle surface morphology, and metal dispersion on the support are all affected by heat treatment. The activity and durability can be achieved from the thermal treatment subsequent to the development of Pt nanoparticles and its alloys [72]. Another study reported, Jalan et al. [73], that thermal treatment of Pt catalyst can slow down the dissolution rate and eventually decrease the initial surface area. The thermal treatment also affects the fundamental characteristics of the catalyst and its support, whereby the number of catalytic sites, the dispersion of catalyst particle on the support, the ratio of catalyst on the support, as well as the acidity-basicity properties of the support [71, 72].

There are several heat treatment technologies, including oven/furnace heating, microwave heating, plasma thermal heating, and ultrasonic spray pyrolysis. The most frequent heating technology is oven/furnace. The heat treatment approach for catalysts that use carbon black as a support for the catalytic metal is known to play two important roles in stability: the loss of oxygen-rich functional groups and graphitization of the carbon support surface. The surface chemistry concerned with carbon blacks exposes new surface states on carbon surfaces with variable oxygen-dominated functionalities that influences the chemical behavior on the surface of carbon support [74].

Firstly, the carbon black (Vulcan XC72R)-supported on Pd-V electrocatalysts were subjected to thermal treatment in 10% H2 under Ar atmosphere at a variety of temperatures. The morphological features of the heat-treated particle were studied, and it was discovered that particle size grew with increasing temperature. Heat treatment was also found to boost the electrocatalytic activity of the catalyst [75].

2. Methodology

2.1 Preparation of binary catalyst PtRu/MWCNT

The PtRu bimetallic nanoparticle catalysts were supported onto the walls of MWCNTs by firstly using the modified polyol reduction method before heat treatment.

2.1.1 Modified polyol method

Catalysts were synthesized using the processes described by Jeng et al. [76] and L. Khotseng et al. [77] for the production of platinum-based binary catalysts, with minor modifications. Ethylene glycol (EG), which works as both a reducing agent and a stabilizer [78], was chosen due to its inherent benefits, which include homogenous distribution and tiny particle size [78, 79].

In a 1:1 ratio, 75 mg of MWCNTs were mixed with 60 ml of ethylene glycol (EG). The mixture was sonicated for 15 minutes and combined at room temperature for 30 minutes to create a homogenous paste. PtRu (1,2 atomic ratio) was synthesized in a separate beaker by adding and dissolving 0.0772 mmol of H2PtCl6H2O as Pt and Ru precursor salts, respectively, in a 3:1 ethylene glycol ratio and stirring at room temperature to obtain a homogeneous mixture. The MWCNT/EG paste was mixed with salt/EG solution to adjust the pH until it reached approximately 3.6 by adding dropwise fraction amounts of 4 M NaOH. To help the metal salts adhere to the surface of the MWCNTs, the resulting mixture was subjected to an hour of vigorous stirring at high speed. The mixture was then refluxed at 160°C for three hours with a constant nitrogen flow, after which it was allowed to cool before being rinsed with ultra-pure

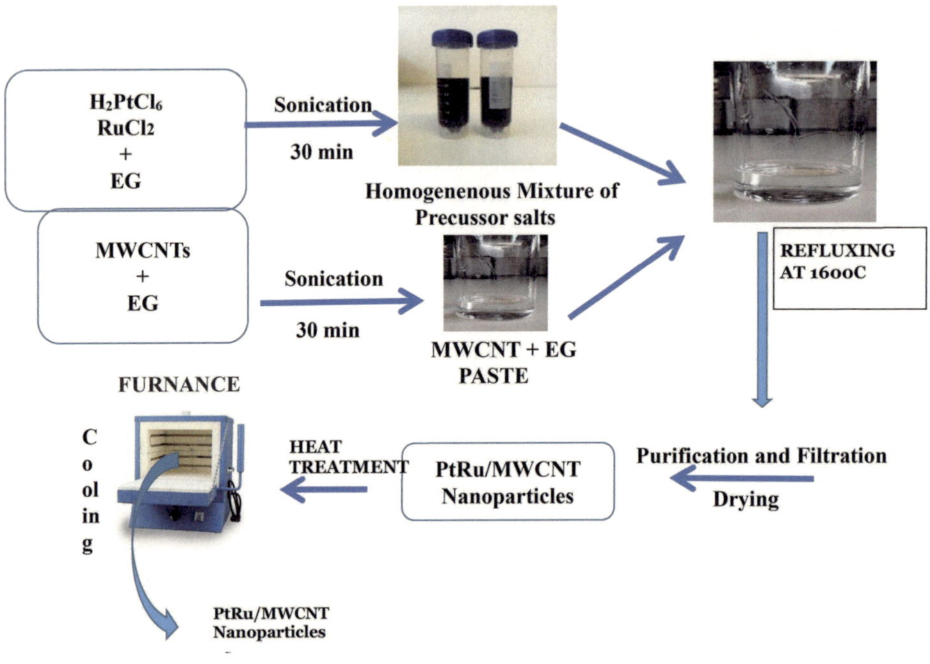

Figure 2.
Schematic diagram for the heat treated PtRu/MWCNT produced by modified polyol method of catalyst preparation.

water. Finally, the mixture was filtered, and the remaining material (PtRu/MWCNT nanoparticle) was oven-dried at 60°C for an extended period of time to produce PtRu/MWCNT catalyst.

2.1.2 Heat treatment of the catalysts

The improved polyol technique was used to produce PtRu/MWCNT inary catalysts, which were then heated under temperature control. Heat treatment has an impact on the shape, activity, and even distribution of the catalysts on the catalytic support. The desired temperature had already been set in the tube furnace. In the middle of the tube furnace, a boat made of alumina held the catalyst. While nitrogen flowed at a rate of 5 ml/min, the catalyst was heated for 3 hours at three target temperatures of 2500C, 3500C, and 4500C. After the heating period was finished, the catalyst was taken out of the furnace and the tube was cooled while the nitrogen gas flowed continuously at a rate of 5 ml/min (**Figure 2**).

3. Results and discussion

3.1 Structural characterization of heat treated PtRu/MWCNT @450°C, PtRu/MWCNT @350°C and PtRu/MWCNT @250°C electrocatalysts

Before the electrocatalysts underwent heat treatment, the metal loading was assessed using energy-dispersive X-ray spectroscopy (EDS). The modified polyol technique was used to manufacture the binary catalyst PtRu/MWCNT, which contains 5.80 percent ruthenium and 2.35 percent platinum. The microstructure of the produced electrocatalysts was evaluated by HR-TEM and XRD, and all catalysts have equivalent atomic ratios.

Scherrer's equation was used to calculate the crystallite sizes of the electrocatalysts, and high-resolution transmission electron microscopy (HR-TEM) was used to measure the particle sizes. Using Debye-Scherrer's equation, K/Cos, where K is the Scherrer constant, = 0.9, the X-ray wavelength is 0.154 nm, and (2), the width of the diffraction peak (rad), one may determine the size of the metal particles' crystals. Using Image J, (a program created by the National Institutes of Health and the Laboratory for Optical and Computational Instrumentation, LOCI, University of Wisconsin, Madison, WI, USA), the particle size determined by HR-TEM was obtained over various regions for each electrocatalyst (**Table 1**).

The face-centered cubic (fcc)crystalline structures are present in all PtRu/CNT electrocatalysts heat treated at 250°C, 350°C, and 450°C, with PtRu/CNT@450deg. having the smallest crystalline size, according to XRD. The X-ray diffraction (XRD) pattern of the Bragg's angles of 39.760, 46.20, 67.40, 81.30, and 85.70 correspond to (111), (200), (220), and (311) and (222), respectively, and is an exact match to the X-ray diffraction pattern of the Pt catalyst. The PtRu crystals' bimetallic architecture was affected by the heat treatment, according to the Bragg angles.

Figure 3 shows the HRTEM micrographs of PtRu electrocatalysts on MWCNT heat treated at 250°C, 350°C and 450°C respectively Catalyst nanoparticle are the dark dots. Multi-walled carbon nanotubes (MWCNTs) support are the large tube-like particles seen, with diameter of about 20 nm. The PtRu nanoparticles got more agglomerated on the MWCNT as temperature increases indicating that the heat treatment made the particles of the electrocatalyst to get more closely packed (**Figure 4**).

Electrocatalysts	Crystalline size(nm)	Particle size(nm)	ECSA
PtRu/MWCNT 250^0	1.178	1.005	5.098m^2/g
PtRu/MWCNT 350^0	0.596	0.616	3.922m^2/g
PtRu/MWCNT 450^0	0.595	0.600	16.714m^2/g

Table 1.
Properties of the Pt-Ru/MWCNT@ 250°C, Pt-Ru/MWCNT@ 350°C and Pt-Ru/MWCNT@ 450°C catalysts.

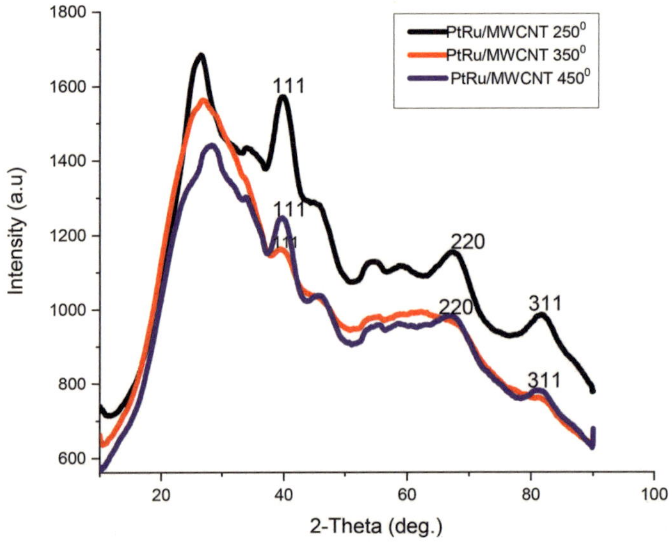

Figure 3.
XRD spectra of PtRu/MWCNT@250°C, PtRu/MWCNT@350°C and PtRu/MWCNT@450°C electrocatalysts supported on MWCNTs.

3.2 Electrocatalytic activity of PtRu/MWCNT electrocatalysts at 2500, 3500, and 4500 degrees Celsius

Cyclic voltammetry was used to initially assess the electrochemical activity of the produced catalysts in a 0.5 M HClO4 solution. The adsorption peaks for the various catalysts could be seen from the cyclic voltammetry of the electrocatalysts that had been synthesized. Eq. (1) was used to calculate the electro-active surface area of the catalysts using the peak area of the adsorption peak of the electrocatalysts in cyclic voltammetry [80].

$$ECSA = \frac{Q}{210\mu C\,/\,cm2.m.Ag} \quad (1)$$

Ag represents the geometric surface area of the electrode (5 mm in diameter), Q represents the charge from the adsorption peak in Coulomb taking within the negative potential region −0.2 V to 0.08 V in the forward scan, and 210 Ccm-2 represents the charge of full coverage for clean polycrystalline Pt monolayer [81].

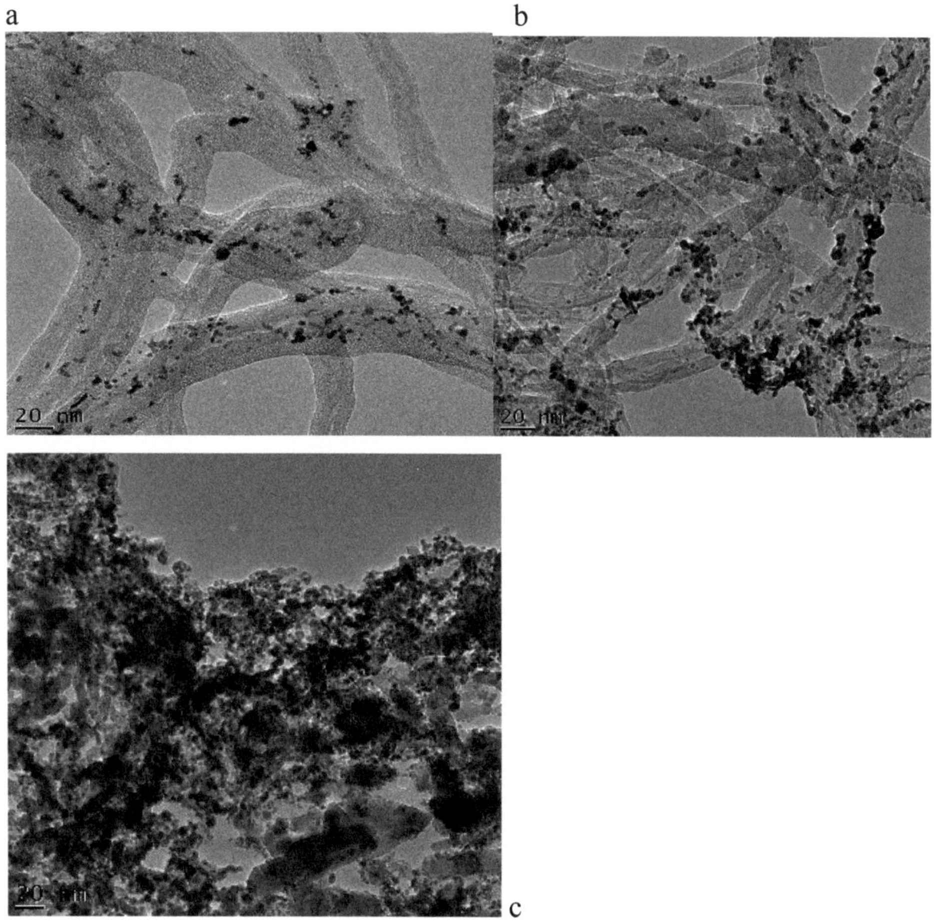

Figure 4.
HR-TEM micrographs of a) PtRu/MWCNT @250°C, b) PtRu/MWCNT @350°C) PtRu/MWCNT @450°C.

The obtained ECSA values are 16.71m^2/g for PtRu/MWCNT@450°C, 3.92m^2/g for PtRu/MWCNT@350°C and 5.098m^2/g for PtRu/MWCNT@250°C. Higher ECSA value of 16.710m^2/g for PtRu/MWCNT@450°C can be attributed to its improved alloying with lowest particle size value of 0.600 nm (**Figure 5**).

3.3 Methanol oxidation on PtRu/MWCNT@2500C, PtRu/MWCNT@3500C, and PtRu/MWCNT@4500C electrocatalysts in 0.5 M Perchloric acid solution at 30mVs^{-1} scan rate

Table 2 summarizes the electrocatalytic activity towards methanol oxidation. By comparing the properties of the CVs, it was discovered that increasing the temperature in the metal alloys significantly increased the catalytic activity for methanol electrooxidation. To begin, the onset potentials (measures of catalytic activity) of methanol oxidation for the heat treated PtRu/MWCNT catalyst at 2500C were lower than for other electrocatalysts. The onset potential positions are as follows: PtRu/MWCNT@450°C > PtRu/MWCNT@350°C > PtRuW/MWCNT@250°C. The forward peak current densities (a measure of maximal catalyst performance) of the binary

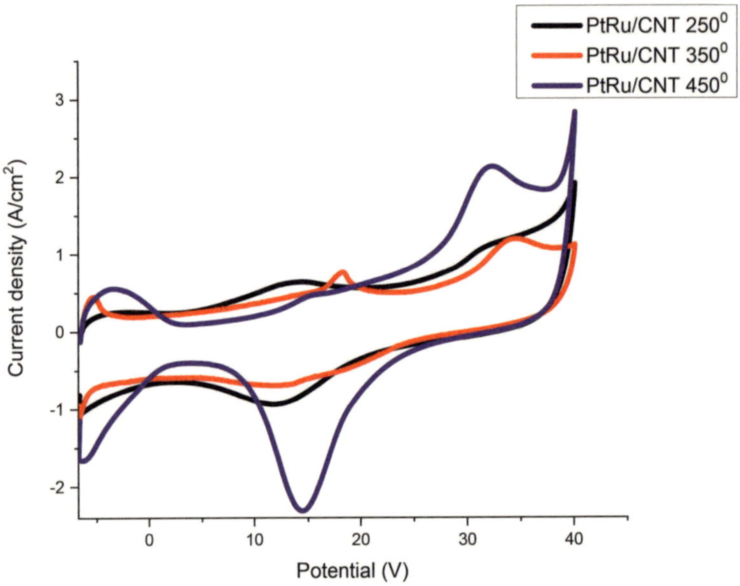

Figure 5.
Cyclic voltammograms of PtRu/MWCNT@250°C, PtRu/MWCNT@350°C and PtRu/MWCNT@450°C electrocatalysts in N_2 saturated 0.5 M Perchloric acid, $HClO_4$ at a scan rate of 30mVs^{-1}.

Electrocatalysts	Onset-Potential(V)	Current Density(mA/cm^2)	Mass Activity {A/g]
PtRu/MWCNT 250^0	0.0841	1.343	13.17
PtRu/MWCNT 350^0	−0.194	0.602	5.90
PtRu/MWCNT 450^0	−0.198	1.987	19.48

Table 2.
Comparison of the electrocatalytic activity of methanol oxidation catalysts.

catalyst PtRu/MWCNT were in the following order: PtRu/MWCNT@450°C > PtRu/MWCNT@2500C > PtRuW/MWCNT@350°C.

Thus, the binary catalyst PtRu/MWCNT@450°C exhibited best electrochemical performance in terms of the highest forward peak current density followed PtRu/MWCNT@250°C (**Figure 6**).

Several researchers thermally treated Pt-based catalysts. Valisi et al. [68] used varied temperatures to heat treat PtCo/C, PtNi/C, PtCu/C, and PtFe/C. PtCu/C at 250°C showed the highest catalytic mass activity in the research. Makodo Uchida et al. [82] used a heat-treated PtRu catalyst as well. The heat treatment is not only favored for the significant durability of the Pt-Ru catalyst in air at 370 degrees, consequently improves methanol oxidation. This process was performed at an initial overpotential of 340 mV vs. NHE at 60 mA cm^{-2}.

We report herein on heat treated bimetallic PtRu/MWCNT as given in **Table 3** below.

PtRu/MWCNT at 450°C showed better stability and kinetics towards the methanol oxidati9n reaction indicating improved electroactivities at higher heat treatment (**Figure 7**).

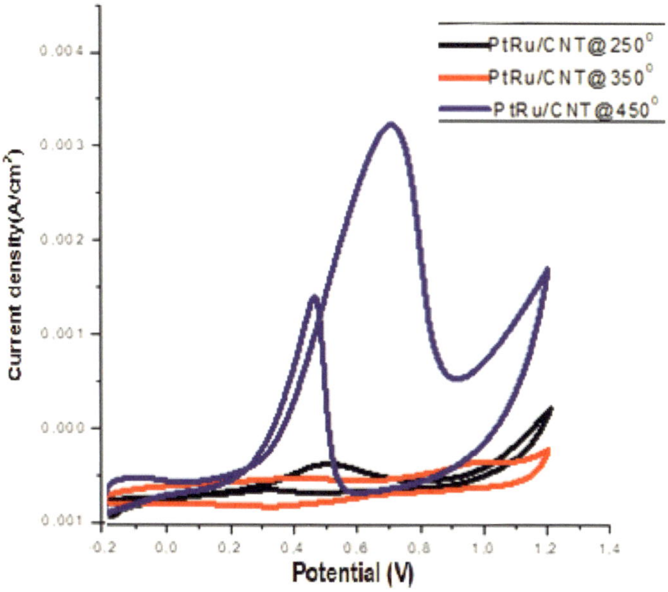

Figure 6.
Cyclic voltammograms of PtRu/MWCNT@250°C, PtRu/MWCNT@350°C and PtRu/MWCNT@450°C, purged with N_2, 2 M methanol, $HClO_4$, saturated 0.5 M Perchloric acid, at a scan rate of 30mVs^{-1}.

Electrocatalysts	Crystalline size(nm)	Particle size(nm)	Atomic ratio determined by EDX	Mass activity (A g − 1 PtRu)
PtRu/MWCNT 250°C	1.178	1.005	Pt62.0:Ru38	13.17
PtRu/MWCNT 350°C	0.596	0.616	Pt68.13:Ru31.87	5.90
PtRu/MWCNT 450°C	0.595	0.600	Pt65.62:Ru34.38	19.48

Table 3.
Different electrocatalysts' compositions, crystalline and particle sizes, and catalytic activity.

Electrochemical impedance spectroscopy technique was used to investigate the catalytic reaction kinetics for the methanol oxidation on the anodic PtRu/MWCNT@250°C, PtRu/MWCNT@350°C and PtRu/MWCNT@250°C electrocatalysts surfaces. The charge transfer resistance, Rct values using Equivalent Circuit fitting were 0.151 kΩ, 2.04 kΩ and 11.31 kΩ for PtRu/MWCNT@450°C, PtRu/MWCNT@350°C and PtRu/MWCNT@250°C respectively indicating that PtRu/MWCNT@450°C exhibited best kinetics towards the methanol electrooxidation with the best conductivity to flow of electric current (**Tables 4** and **5**).

PtRu/MWCNT heat treated at 450^0 gave best kinetics as it offered least resistance to the flow of current with charge transfer resistance value of 0.151 kΩ followed by PtRu/MWCNT heat treated at 350°C.

PtRu/MWCNT heat treated at 450^0 gave best kinetics as it offered least resistance to the flow of current with charge transfer resistance value of 0.151 kΩ followed by PtRu/MWCNT heat treated at 350°C (**Figure 8**).

Stability is critical for electrocatalysts to be employed efficiently in DMFCs. **Figure 9** shows the chronoamperometry (CA) of PtRu/MWCNT electrocatalysts

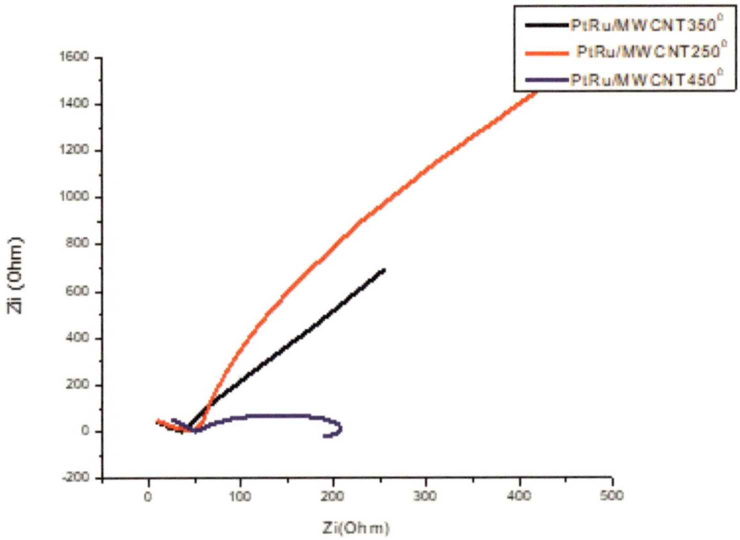

Figure 7.
EIS curves of methanol oxidation on PtRu/MWCNT electrocatalysts prepared from various synthesis routes purged with N_2 in saturated 0.5 M HClO4 and 0.2 M methanol.

Electrocatalysts	Charge transfer resistance (R_{CT})
PtRu/MWCNT 250°C	11.31 kΩ
PtRu/MWCNT 350°C	2.040 kΩ
PtRu/MWCNT 450°C	0.151 kΩ

Table 4.
Electrocatalysts with PtRu/MWCNT charge transfer resistance.

Electrocatalysts	Charge transfer resistance (R_{CT})
PtRu/MWCNT 250°C	11.31 kΩ
PtRu/MWCNT 350°C	2.040 kΩ
PtRu/MWCNT 450°C	0.151 kΩ

Table 5.
Charge transfer resistance of PtRu/MWCNT electrocatalysts.

on MWCNT support in N2 saturated 0.5 M HClO4 with 2.0 M methanol. The importance was sought after 1800 seconds on various catalysts. The current density studies were revealed on chronoamperometry analysis, an initial observation was rapid decay with time (I proportional to t-1/2). The rate of inhibition of the electrodes by the products of the methanol oxidation reaction may decrease over time. When manufactured catalysts, PtRu/MWCNT binary catalysts heat treated at different temperatures are compared, the order of stability of the electroacatalysts to methanol electrooxidation is as follows. PtRu/MWCNT at 4500C > PtRu/MWCNT at 2500C > PtRu/MWCNT at 3500C based on current density values of 0.284 mA/cm2 (**Table 6**).

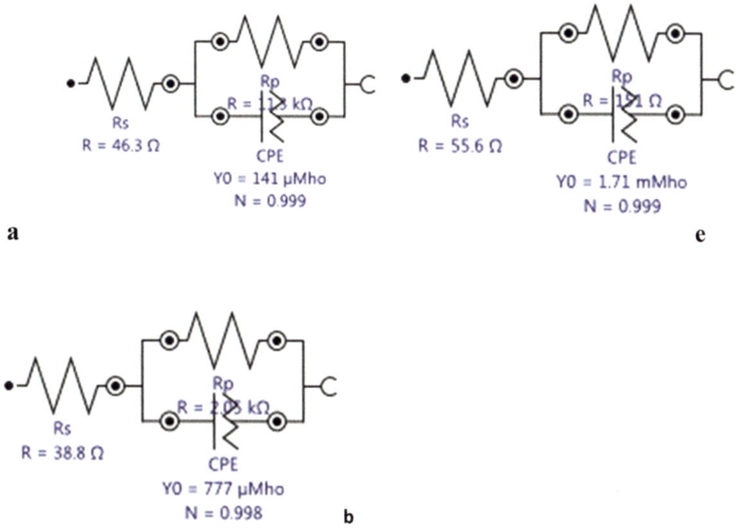

Figure 8.
Equivalent circuits for methanol oxidation electrochemical impedance spectroscopy on (a) PtRu/MWCNT@4500C, (b) PtRu/MWCNT@3500C, and (c) PtRu/MWCNT@2500C electrocatalysts in N2 saturated 0.5 M HClO4 and 0.2 M methanol.

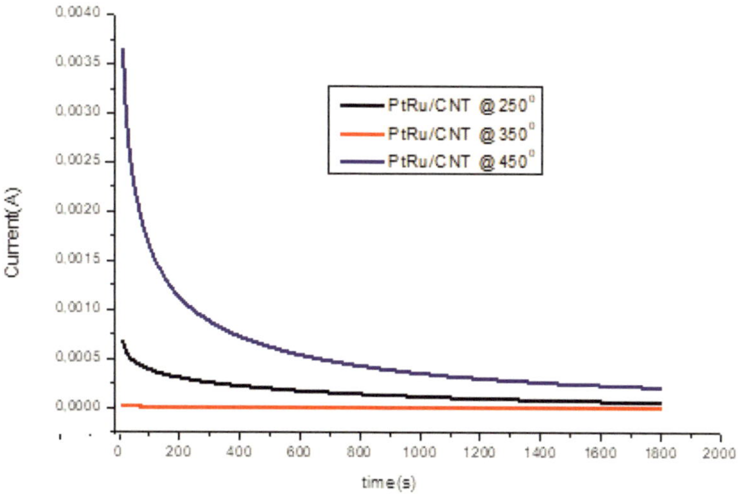

Figure 9.
The Chronoamperometry curves of methanol oxidation on PtRu/MWCNT electrocatalysts in 0.5 M $HClO_4$ and 2.0 M CH_3OH.

Electrocatalysts	Current Density (mA/cm^2)
PtRu/MWCNT 250^0	0.284
PtRu/MWCNT 350^0	0.0329
PtRu/MWCNT 450^0	1.296

Table 6.
Current density values from the chronoamperometry curve for the stability test of the PtRu/MWCNT electrocatalysts.

4. Conclusion

In this study, PtRu supported on MWCNT was successfully synthesized using the modified polyol approach and heat treated at 2500C. 3500 and 4500 degrees Celsius, respectively. The synthesized electrocatalysts had crystalline diameters of 0.595–1.178 nm and average particle sizes of 0.699–1.005 nm, according to XRD and HRTEM studies. The PtRu alloy phase, according to XRD studies, is an excellent match for the Pt catalyst structure. The PtRu/MWCNT electrocatalyst heat-treated at 4500 C was shown to have higher electrocatalytic activity for methanol oxidation, with a mass activity of 19.48 A/g, than other PtRu electrocatalysts on MWCNT support. When compared to other electrocatalysts, the 4500C PtRu/MWCNT electrocatalyst displayed the maximum current density for methanol oxidation. This is because it participates most actively in the anode oxidation reaction and has the lowest particle size. The PtRu/MWCNT electrocatalysts heated to 4500C showed faster electrochemical reaction kinetics than those heated to 2500C and 3500C with a charge transfer resistance value of 0.151 k Ω, according to the EIS.

Finally, the temperature 450°C was found to be optimal as the PtRu/MWCNT electrocatalyst heat treated at this temperature is also the most stable electrocatalyst followed by the PtRu/MWCNT electrocatalyst heat treated at 250°C as shown by the chronoamperomerty tests.

Author details

Adebare Nurudeen Adewunmi[1*], Ntalane Sello Seroka[1], Su Huaneng[2] and Khotseng Lindiwe Eudora[1*]

1 Faculty of Natural Sciences, Department of Chemistry, University of the Western Cape, Cape Town, South Africa

2 Institute for Energy Research, Jiangsu University, Zhenjiang, China

*Address all correspondence to: 3875793@myuwc.ac.za and lkhotseng@uwc.ac.za

References

[1] O'hayre R, Cha S-W, Colella W, Prinz FB. Fuel cell fundamentals. John Wiley & Sons; 2016

[2] Elmer T, Worall M, Wu S, Riffat SB. Fuel cell technology for domestic built environment applications: State of-the-art review. Renewable and Sustainable Energy Reviews. 2015;**42**:913-931

[3] Sharaf OZ, Orhan MF. An overview of fuel cell technology: Fundamentals and applications. Renewable and Sustainable Energy Reviews. 2014;**32**:810-853

[4] Fernandez Alvarez, G., 2011. Palladium Based Catalysts for Oxygen Reduction in Polymer Electrolyte Membrane Fuel Cells (Doctoral thesis, Newcastle University). [Accessed online, 28 February 2017]

[5] Lucia U. Overview on fuel cells. Renewable and Sustainable Energy Reviews. 2014;**30**:164-169

[6] Lai J, Ellis MW. Fuel cell power systems and applications. Proceedings of the IEEE. 2017;**105**:2166-2190

[7] Available from: https://cdn.shortpixel.ai/spai/q_lossy+ret_img/https://i2.wp.com/www.electroniclinic.com/wp-conte

[8] U.S. EERE U. Department of Energy's Office of Energy Efficiency and Renewable Energy. Geothermal Electricity Technology Evaluation Model (GETEM); 2016

[9] Antolini E. Palladium in fuel cell catalysis. Energy & Environmental Science. 2009;**2**(9):915-931

[10] Antolini E, Perez J. The renaissance of unsupported nanostructured catalysts for low-temperature fuel cells: From the size to the shape of metal nanostructures. Journal of Materials Science. 2011;**46**:4435-4457

[11] Lu Y, Tu JP, Gu CD, Xia XH, Wang XL, Mao SX. Growth of and methanol electro-oxidation by gold nanowires with high density stacking faults. Journal of Materials Chemistry. 2011;**21**(13):4843-4849

[12] Hogarth M, Ralph T. Catalysis for low temperature fuel cells. Platinum Metals Review. 2002;**46**:146-164

[13] Hamnett A. Mechanism and electrocatalysis in the direct methanol fuel cell. Catalysis Today. 1997;**38**:445-457

[14] Yajima T, Uchida H, Watanabe M. In-situ ATR-FTIR spectroscopic study of electro-oxidation of methanol and adsorbed CO at Pt– Ru alloy. The Journal of Physical Chemistry B. 2004;**108**(8):2654-2659

[15] Ambrosio EP, Francia C, Manzoli M, Penazzi N, Spinelli P. Platinum catalyst supported on mesoporous carbon for PEMFC. International Journal of Hydrogen Energy. 2008;**33**(12):3142-3145

[16] Lin Y, Cui X, Yen CH, Wai CM. PtRu/carbon nanotube nanocomposite synthesized in supercritical fluid: A novel electrocatalyst for direct methanol fuel cells. Langmuir. 2005;**21**(24):11474-11479

[17] Kim YI et al. Electrocatalytic properties of carbon nanofiber web-supported nanocrystalline Pt catalyst as applied to direct methanol fuel cell. International Journal of Electrochemical Science. 2009;**4**:1548-1559

[18] Maiyalagan T. Silicotungstic acid stabilized Pt–Ru nanoparticles supported

on carbon nanofibers electrodes for methanol oxidation. International Journal of Hydrogen Energy. 2009;**34**:2874

[19] Tiwari JN, Tiwari RN, Singh G, Kim KS. Recent progress in the development of anode and cathode catalysts for direct methanol fuel cells. Nano Energy. 2013;**2**(5):553-578

[20] Neto AO, Dias RR, Tusi MM, Linardi M, Spinacé EV. Electro-oxidation of methanol and ethanol using PtRu/C, PtSn/C and PtSnRu/C electrocatalysts prepared by an alcohol-reduction process. Journal of Power Sources. 2007;**166**(1):87-91

[21] Hassan HB. Electrodeposited Pt and Pt-Sn nanoparticles on Ti as anodes for direct methanol fuel cells. Journal of Fuel Chemistry and Technology. 2009;**37**(3):346-354

[22] Qi Z, Geng H, Wang X, Zhao C, Ji H, Zhang C, et al. Novel nanocrystalline PdNi alloy catalyst for methanol and ethanol electro-oxidation in alkaline media. Journal of Power Sources. 2011;**196**(14):5823-5828

[23] Lee SA, Park KW, Choi JH, Kwon BK, Sung YE. Nanoparticle synthesis and electrocatalytic activity of Pt alloys for direct methanol fuel cells. Journal of the Electrochemical Society. 2002;**149**(10):A1299

[24] Morante-Catacora TY, Ishikawa Y, Cabrera CR. Sequential electrodeposition of Mo at Pt and PtRu methanol oxidation catalyst particles on HOPG surfaces. Journal of Electroanalytical Chemistry. 2008;**621**(1):103-112

[25] Chen CS, Pan FM. Electrocatalytic activity of Pt nanoparticles deposited on porous TiO2 supports toward methanol oxidation. Applied Catalysis B: Environmental. 2009;**91**(3-4):663-669

[26] Xing L, Jia J, Wang Y, Zhang B, Dong S. Pt modified TiO2 nanotubes electrode: Preparation and electrocatalytic application for methanol oxidation. International journal of hydrogen energy. 2010;**35**(22):12169-12173

[27] Huang J, Yang H, Huang Q, Tang Y, Lu T, Akins DL. Methanol oxidation on carbon-supported Pt-Os bimetallic nanoparticle electrocatalysts. Journal of the Electrochemical Society. 2004;**151**(11):A1810

[28] Kang Y, Murray CB. Synthesis and electrocatalytic properties of cubic Mn–Pt nanocrystals (nanocubes). Journal of the American Chemical Society. 2010;**132**(22):7568-7569

[29] Kordesch K, Simader G. Fuel cells and their applications. 1996

[30] Léger J-M. Mechanistic aspects of methanol oxidation on platinum-based electrocatalysts. Journal of Applied Electrochemistry. 2001;**31**:767-771

[31] Lim C, Wang C. Development of high-power electrodes for a liquid-feed direct methanol fuel cell. Journal of Power Sources. 2003;**113**:145-150

[32] Papoutsis A, Léger JM, Lamy C. Study of the kinetics of adsorption and electro-oxidation of MeOH on Pt (100) in an acid medium by programmed potential voltammetry. Journal of Electroanalytical Chemistry. 1993;**359**(1-2):141-160

[33] Herrero E, Franaszczuk K, Wieckowski A. Electrochemistry of methanol at low index crystal planes of platinum: An integrated voltammetric and chronoamperometric study. The Journal of Physical Chemistry. 1994;**98**(19):5074-5083

[34] Biswas PC, Nodasaka Y, Enyo M. Electrocatalytic activities of

graphite-supported platinum electrodes for methanol electrooxidation. Journal of applied electrochemistry. 1996;**26**:30-35

[35] Burstein GT, Barnett CJ, Kucernak AR, Williams KR. Aspects of the anodic oxidation of methanol. Catalysis Today. 1997;**38**(4):425-437

[36] Lin WF, Wang JT, Savinell RF. On-line FTIR spectroscopic investigations of methanol oxidation in a direct methanol fuel cell. Journal of the Electrochemical Society. 1997;**144**(6):1917

[37] Zhu Y et al. Attenuated total reflection– Fourier transform infrared study of methanol oxidation on sputtered Pt film electrode. Langmuir. 2001;**17**(1):146-154

[38] Christensen PA, Hamnett A, Troughton GL. The role of morphology in the methanol electro-oxidation reaction. Journal of Electroanalytical Chemistry. 1993;**362**(1-2):207-218

[39] Pletcher D, Solis V. The effect of experimental parameters on the rate and mechanism of oxidation of methanol at a platinum anode in aqueous acid. Electrochimica Acta. 1982;**27**(6):775-782

[40] Yahikozawa K et al. Electrocatalytic properties of ultrafine platinum particles for oxidation of methanol and formic acid in aqueous solutions. Electrochimica Acta. 1991;**36**(5-6):973-978

[41] Kinoshita K. Particle size effects for oxygen reduction on highly dispersed platinum in acid electrolytes. Journal of the Electrochemical Society. 1990;**137**(3):845

[42] Takasu Y, Fujii Y, Matsuda Y. Size effects of small platinum particles on the electrocatalytic oxidation of methanol. Bulletin of the Chemical Society of Japan. 1986;**59**(12):3973-3974

[43] Burke LD, Casey JK. Hydrous oxide formation on platinized platinum electrodes and its relevance to oxygen gas reduction. Berichte der Bunsengesellschaft für Physikalische Chemie. 1990;**94**(9):931-937

[44] Watanabe M, Saegusa S, Stonehart P. High platinum electrocatalyst utilizations for direct methanol oxidation. Journal of Electroanalytical Chemistry and Interfacial Electrochemistry. 1989;**271**(1-2):213-220

[45] Frelink T, Visscher W, Van Veen JAR. Particle size effect of carbon-supported platinum catalysts for the electrooxidation of methanol. Journal of Electroanalytical Chemistry. 1995;**382**(1-2):65-72

[46] Kennedy BJ, Hamnett A. Oxide formation and reactivity for methanol oxidation on platinised carbon anodes. Journal of Electroanalytical Chemistry and Interfacial Electrochemistry. 1990;**283**(1-2):271-285

[47] Leung LWH, Michael J. Weaver. Influence of adsorbed carbon monoxide on electrocatalytic oxidation of simple organic molecules at platinum and palladium electrodes in acidic solution: A survey using real-time FTIR spectroscopy. Langmuir. 1990;**6**(2):323-333

[48] Marković NM, Gasteiger HA, Ross PN Jr, Jiang X, Villegas I, Weaver MJ. Electro-oxidation mechanisms of methanol and formic acid on Pt-Ru alloy surfaces. Electrochimica Acta. 1995;**40**:91-98

[49] Iwasita-Vielstich T. In: Gerischer H, Tobias CW, editors. Advances in Electrochemical Science and

Engineering. 1st ed. Weinheim: VCH; 1990. pp. 127-170

[50] Ma C, Chen Z, Zhao F. Synthesis of ultrafine mesoporous tungsten carbide by high-energy ball milling and its Electrocatalytic activity for methanol oxidation. Chinese Journal of Chemistry. 2011;**29**(4):611-616

[51] Zhong X, Zhang X, Sun X, Liu B, Kuang Y, Chen J. Pt and Pt-Ru nanoparticles dispersed on Ethylenediamine grafted carbon nanotubes as new Electrocatalysts: Preparation and Electrocatalytic properties for ethanol Electrooxidation. Chinese Journal of Chemistry. 2009;**27**(1):56-62

[52] Martínez-Huerta MV, Rojas S, De La Fuente JG, Terreros P, Pena MA, Fierro JL. Effect of Ni addition over PtRu/C based electrocatalysts for fuel cell applications. Applied Catalysis B: Environmental. 2006;**69**(1-2):75-84

[53] Pasupathi S, Tricoli V. Effect of third metal on the electrocatalytic activity of PtRu/Vulcan for methanol electro-oxidation. Journal of Solid State Electrochemistry. 2008;**12**:1093-1100

[54] Huang T, Liu J, Li R, Cai W, Yu A. A novel route for preparation of PtRuMe (me= Fe, Co, Ni) and their catalytic performance for methanol electrooxidation. Electrochemistry Communications. 2009;**11**(3):643-646

[55] Gurau B et al. Structural and electrochemical characterization of binary, ternary, and quaternary platinum alloy catalysts for methanol electro-oxidation. The Journal of Physical Chemistry B. 1998;**102**(49):9997-10003

[56] Choi WC, Kim JD, Woo SI. Quaternary Pt-based electrocatalyst for methanol oxidation by combinatorial electrochemistry. Catalysis Today. 2002;**74**(3-4):235-240

[57] Arico AS et al. Characterization of direct methanol fuel cell components by electron microscopy and X-ray microchemical analysis. Materials Chemistry and Physics. 1997;**47**(2-3):257-262

[58] Aramata A, Masuda M. Platinum alloy electrodes bonded to solid polymer electrolyte for enhancement of methanol electro-oxidation and its reaction mechanism. Journal of the Electrochemical Society. 1991;**138**(7):1949

[59] Greenwood NN, Earnshaw A. Chemistry of the Elements. 2nd ed. Oxford: Butterworth-Heinemann; 1997

[60] Debe MK. Electrocatalyst approaches and challenges for automotive fuel cells. Nature. 2012;**486**(7401):43-51

[61] Steele BC, Heinzel A. Materials for fuel-cell technologies. Nature. 2001;**414**(6861):345-352

[62] Kim HJ, Ahn YD, Kim J, Kim KS, Jeong YU, Hong JW, et al. Surface elemental distribution effect of Pt-Pb hexagonal nanoplates for electrocatalytic methanol oxidation reaction. Chinese Journal of Catalysis. 2020;**41**(5):813-819

[63] Schultz MG, Diehl T, Brasseur GP, Zittel W. Air pollution and climate-forcing impacts of a global hydrogen economy. Science. 2003;**302**(5645):624-627

[64] Zhao X, Yin M, Ma L, Liang L, Liu C, Liao J, et al. Recent advances in catalysts for direct methanol fuel cells. Energy & Environmental Science. 2011;**4**(8):2736-2753

[65] Cruz JC, Baglio V, Siracusano S, Ornelas R, Arriaga LG, Antonucci V,

Aricò AS. Nanosized Pt/IrO2 electrocatalyst prepared by modified polyol method for application as dual function oxygen electrode in unitized regenerative fuel cells. International journal of hydrogen energy. 1 Apr 2012;**37**(7):5508-5517

[66] Baglio V, Di Blasi A, D'Urso C, Antonucci V, Aricò AS, Ornelas R, et al. Development of Pt and Pt–Fe catalysts supported on multiwalled carbon nanotubes for oxygen reduction in direct methanol fuel cells. Journal of the Electrochemical Society. 2008;**155**(8):B829

[67] Bezerra CW, Zhang L, Liu H, Lee K, Marques AL, Marques EP, et al. A review of heat-treatment effects on activity and stability of PEM fuel cell catalysts for oxygen reduction reaction. Journal of Power Sources. 2007;**173**(2):891-908

[68] Valisi AN, Maiyalagan T, Khotseng L, Linkov V, Pasupathi S. Effects of heat treatment on the catalytic activity and methanol tolerance of carbon-supported platinum alloys. Electrocatalysis. 2012;**3**(2):108-118

[69] Li X, Liu J, Huang Q, Vogel W, Akins DL, Yang H. Effect of heat treatment on stability of gold particle modified carbon supported Pt–Ru anode catalysts for a direct methanol fuel cell. Electrochimica Acta. 2010;**56**(1):278-284

[70] Chaisubanan N, Maniwan W, Hunsom M. Effect of heat-treatment on the performance of PtM/C (M= Cr, Pd, Co) catalysts towards the oxygen reduction reaction in PEM fuel cell. Energy. 2017;**127**:454-461

[71] Kamarudin SK, Achmad F, Daud WRW. Overview on the application of direct methanol fuel cell (DMFC) for portable electronic devices. International Journal of Hydrogen Energy. 2009;**34**:6902-6916

[72] Gerwith AA, Thorum MS. Electroreduction of dioxygen for fuel-cell applications: Materials and challenges. Inorganic Chemistry. 2010;**49**:355

[73] Janssen MMP, Moolhuysen J. Platinum—tin catalysts for methanol fuel cells prepared by a novel immersion technique, by electrocodeposition and by alloying. Electrochimica Acta. 1976;**21**:861

[74] Wang H, Li H, Yan X. PEM Fuel Cell Failure Mode Analysis. 2011. p. 53

[75] Ang S, Walsh AD. Palladium–vanadium alloy electrocatalysts for oxygen reduction: Effect of heat treatment on electrocatalytic activity and stability. Applied Catalysis B: Environmental. 2010;**98**:49-56

[76] Jeng KT, Chien CC, Hsu NY, Yern SC, Chiou SD, Lin SH, et al. Performance of direct methanol fuel cell using carbon nanotube-supported Pt–Ru anode catalyst with controlled composition. Journal of Power Sources. 2006;**160**:97

[77] Khotseng L, Bangisa A, Modibedi RM, Linkov V. Electrochemical evaluation of Pt-based binary catalysts on various supports for the direct methanol fuel cell. Electrocatalysis. 2016;7:1-12

[78] Liu H, Song C, Zhang L, Zhang J, Wang H, Wilkinson DP. A review of anode catalysis in the direct methanol fuel cell. Journal of Power Sources. 2006;**155**(2):95-110

[79] Kumar SS, Hidyatai N, Herrero JS, Irusta S, Scott K. Efficient tuning of the Pt nano-particle mono-dispersion on Vulcan XC-72R by selective pre-treatment and electrochemical

evaluation of hydrogen oxidation and oxygen reduction reactions. International Journal of Hydrogen Energy. 2011;**36**(9):5453-5465

[80] Garsany Y, Baturina OA, Swider-Lyons KE, Kocha SS. Experimental methods for quantifying the activity of platinum electrocatalysts for the oxygen reduction reaction. Analytical Chemistry. 2010;**82**:6321-6328

[81] Vielstich W, Gasteiger HA, Lamm A. Handbook of Fuel Cells Fundamentals, Technology and Applications. Vol. 2. New York, NY: Wiley; 2003. p. 316. ISBN 0 471 49926 9; TRN: GB0400273

[82] Uchida M, Aoyama Y, Tanabe M, Yanagihara N, Eda N, Ohta A. Influences of both carbon supports and heat-treatment of supported catalyst on electrochemical oxidation of methanol. Journal of the Electrochemical Society. 1995;**142**(8):2572

Chapter 6

Catalysis with Ruthenium for Sustainable Carbon Cycles

Thomas Ernst Müller

Abstract

Nestled between the noble and non-noble metals in the periodic table, ruthenium, one of the transition metals, offers a combination of intriguing properties. Due to its variable oxidation states and its ability to form complexes with various Lewis base compounds, ruthenium, has been widely used in the field of catalysis. Its application has led to groundbreaking breakthroughs in a variety of chemical transformations and has attracted considerable attention in both academic research and industrial applications. Ruthenium catalysis is a dynamic and rapidly evolving field, with ongoing efforts to further advance the efficiency and selectivity of these catalysts. Importantly, in the context of sustainability, ruthenium-based catalysts play an important role in promoting green chemistry practices. Because ruthenium catalysts are highly efficient, only small amounts of the element need to be used. Recovery rates at the end of catalyst life are typically very high, minimizing the need to mine fresh ore. The use of ruthenium catalysts promotes the utilization of renewable resources in various chemical transformations, is at the heart of the realization of new energy-related processes, and by enabling efficient and highly selective chemical transformations reduces waste and harmful emissions. These aspects reinforce the metal's importance in the quest for a more sustainable future.

Keywords: ruthenium, catalysis, hydrogenation, sustainability, carbon cycles

1. Introduction

Sustainable chemistry is increasingly becoming an essential tool in the quest for greener processes and materials. It plays a critical role in conserving resources, reducing pollution, and ensuring the long-term health and survival of our planet. In this context, the element ruthenium is gaining prominence as a homogeneous or heterogeneous catalyst due to its intriguing combination of properties and versatile applications in sustainable chemistry. These include hydrogenation, reductive amination, hydroformylation, and other transformations in sustainable chemistry.

Ruthenium, which occupies a unique position between noble metals and nonnoble metals, possesses distinctive properties that will be discussed in terms of structure-property relationships in this chapter. A major focus will be on how ruthenium catalysts can be tuned for the creation of anthropogenic carbon cycles. One aspect we'll explore is the role of the interaction of metallic ruthenium particles with the support.

Another fascinating aspect is the influence of bimetallic catalysts, which can significantly alter the reaction pathways compared to their monometallic counterparts. In addition, this chapter will cover a variety of industrial applications of heterogeneous ruthenium catalysts. These include the reversible hydrogenation of liquid organic hydrogen carriers (LOHC) and ammonia synthesis in the context of hydrogen storage and lignin utilization, highlighting the importance of ruthenium in the broader context of sustainable chemistry.

1.1 Homogeneous and heterogeneous catalysis with ruthenium

The use of ruthenium catalysts has spanned numerous areas of catalysis, including, but not limited to, the following areas:

- **Hydrogenation Reactions**: The ability of ruthenium catalysts to facilitate hydrogenation reactions has been exploited in the synthesis of a wide range of compounds, with applications ranging from pharmaceuticals to agrochemicals. Ruthenium catalysts, thereby, accommodate a broad range of substrates:

    - Ruthenium catalysts can hydrogenate carbon-carbon double bonds in a variety of alkenes, facilitating the production of alkanes. Ruthenium catalysts can selectively hydrogenate carbon-carbon triple bonds in alkynes to alkenes (partial hydrogenation) or to alkanes (complete hydrogenation). Ruthenium catalysts can also facilitate the hydrogenation of aromatic rings, leading to the production of cycloalkanes.

    - Ruthenium catalysts can be used to hydrogenate carbonyl groups in aldehydes and ketones, forming alcohols. Ruthenium catalysts can hydrogenate carboxylic acids, esters, and amides, forming alcohols and amines.

    - Ruthenium catalysts can be used in the hydrogenation of nitro groups, nitriles, and other unsaturated nitrogen moieties forming amines.

    Ruthenium complexes can also act as catalysts in transfer hydrogenation reactions. The Noyori catalysts, which are complexes of ruthenium with chiral ligands, have been used in the asymmetric transfer of hydrogenation of ketones to alcohols. Ruthenium catalysts also effectively catalyze the hydrogenolytic breaking of chemical bonds, mostly oxygen-carbon bonds, in a molecule.

- **Amine Formation**: Ruthenium catalysts are used in the formation of amines in a number of ways. For instance, they facilitate the reductive amination of carbonyl compounds. In this reaction, an aldehyde or ketone is reacted with an amine in the presence of a hydrogen source. The ruthenium catalyst aids in the transfer of hydrogen from the source to the carbonyl group, forming a new amine. Ruthenium catalysts are also used in the borrowing hydrogen methodology (also known as hydrogen autotransfer) to synthesize amines. In this reaction, a ruthenium catalyst facilitates the removal of hydrogen from an alcohol, allowing it to react with an amine to form an imine intermediate. The catalyst then helps return the borrowed hydrogen to this intermediate to form the final amine product.

- **Olefin Metathesis**: Probably the best-known application of ruthenium catalysts is in olefin metathesis [1], a reaction that rearranges the alkene moieties of olefins while maintaining the number of double bonds. This reaction has undergone transformative advances with the introduction of ruthenium-based catalysts, such as the Grubbs catalysts. These catalysts exhibit high stability, are compatible with a wide range of substrates, and tolerate various functional groups. Ruthenium catalysts are also active for various C–C bond formation reactions [2], including cross-coupling reactions, cycloaddition reactions, and cyclization reactions.

- **Oxidation Reactions**: Ruthenium catalysts are also adept at facilitating oxidation reactions, such as alcohol oxidation, olefin epoxidation, and amine oxidation.

1.2 Electrocatalysis with ruthenium

Ruthenium is also widely used as an electrocatalyst. Well-known is the use of ruthenium in anodes for chloralkali electrolysis. The anodes are often constructed by electrochemically depositing a thin layer of ruthenium oxide and other mixed metal oxides on a conductive titanium substrate. This thin layer, typically only a few microns thick, is where the chlorine evolution reaction takes place. One of the primary factors contributing to the high performance of these ruthenium-based anodes is their ability to maintain a low chlorine overpotential. In addition, these anodes exhibit excellent resistance to the highly corrosive environment within the electrolytic cell, including a highly corrosive electrolyte and high current densities. Their high chemical stability and good electronic conductivity make ruthenium-based anodes suitable for long-term use in industrial-scale chloralkali electrolysis. Ruthenium is also a key anode component in water-splitting electrolyzers, which are central to the production of hydrogen as a clean and sustainable energy source. Ruthenium-based catalysts, often mixed with other metals such as iridium or titanium, have demonstrated high efficiency in the oxygen evolution reaction (OER), one of the half-reactions in water splitting. Similarly, ruthenium complexes are being explored for use used as electrocatalysts in organic electrosynthesis, where they allow the transformation of organic compounds under mild conditions with high selectivity. Ruthenium is used in the anode catalyst layer of proton exchange membrane fuel cells (PEMFC), where it catalyzes the oxidation of hydrogen or small organic molecules. It is also used as an electrocatalyst in metal-air batteries. Metal-air batteries combine a metal anode with an oxygen cathode that uses ambient oxygen from the air, separated by an electrolyte. They have attracted considerable interest due to their high theoretical energy densities, making them promising candidates for use in electric vehicles and large-scale grid storage. Ruthenium, with its unique electrochemical properties, has been shown to be an efficient electrocatalyst for the oxygen reduction reaction (ORR) and oxygen evolution reaction (OER) that occur at the cathode in metal-air batteries. ORR occurs during the discharge phase of the battery, while OER occurs during the charge phase. Efficiently catalyzing these reactions is critical to improving the energy efficiency and cycle life of the battery.

1.3 Role of ruthenium in the context of sustainability

Ruthenium catalysts could play an important role in meeting the challenges associated with increased greenhouse gas emissions. An important contribution to mitigating climate change will be the establishment of anthropogenic carbon cycles [3]. Below, we will discuss some of the potential contributions that ruthenium catalysis could make. The focus will be on exploiting the specific advantages that ruthenium catalysts can offer:

i. Ruthenium catalysts often operate at lower temperatures and pressures than other catalysts, thereby reducing energy consumption. In addition, these catalysts can be highly selective, resulting in fewer by-products and reducing the need for energy-intensive purification steps.

ii. Many ruthenium-catalyzed reactions, such as olefin metathesis or hydrogenation reactions, have high atom economy, meaning that a large fraction of the atoms from the starting materials end up in the final product. This reduces by-product formation and improves the overall sustainability of the process.

iii. Ruthenium catalysts can often be recovered and reused, reducing the need for new catalyst production and the need to mine ruthenium ores.

iv. With regard to energy applications, ruthenium complexes have been established for use in dye-sensitized solar cells, a type of thin-film solar cell. This could be an important contribution to more sustainable energy production.

v. Ruthenium catalysts have been investigated for their ability to make use of renewable feedstock from biological sources, such as lignin [4], and CO_2 [5], by reducing it to useful chemicals and fuels. Such applications could be an important component in establishing a circular carbon economy.

In the broader context of sustainability and climate change, the use of ruthenium catalysts in chemical processes could be part of the shift to cleaner and more efficient production methods that minimize the release of CO_2 and even effectively reuse it. It is important to note, however, that this would be only one piece of a much larger puzzle in mitigating human impacts [6] on carbon fluxes through geohabitats [3] and the climate [7].

2. Catalytic transformations with ruthenium catalysts

2.1 Hydrogenation reactions

The strategic deployment of ruthenium catalysts to facilitate hydrogenation reactions holds substantial promise, particularly in energy-related applications. Hydrogenation is a chemical transformation involving the addition of hydrogen (H_2) to a substrate, and it requires catalysts to lower the activation energy and increase the reaction rate to a technically viable scale. Ruthenium catalysts have demonstrated superior efficacy in such roles. In the energy field, ruthenium is being considered for applications in hydrogen storage, where the reversible hydrogenation and dehydrogenation of specific compounds allow for compact and efficient storage of hydrogen. Ruthenium catalysts are also employed in the hydrogenation of carbon dioxide to formic acid derivatives, a promising area for sustainable chemistry. Furthermore, in the production of synthetic fuels or synfuels, ruthenium catalysts are used to hydrogenate carbon dioxide to energy-rich liquid fuels. This conversion is a modified version of the Fischer-Tropsch process, which converts syngas, a mixture of carbon monoxide and hydrogen, into liquid hydrocarbons. These hydrocarbons serve as a potential renewable replacement for conventional petroleum-based fuels. Recent advances extend this concept to use a mixture of carbon dioxide and hydrogen to produce liquid hydrocarbons. With these diverse applications, ruthenium catalysts are paving the way for more sustainable and efficient energy solutions, underscoring their critical role in hydrogenation reactions in the energy sector.

2.1.1 Hydrogen storage

Renewable energy from many primary energy sources, such as wind and solar power, is inherently intermittent in nature. To balance the resulting mismatch between energy supply and demand, suitable energy storage is required. Electricity can be stored only in small quantities in physical form, e.g., in pump storage plants or batteries. An alternative is a storage of renewable energy in chemicals as secondary energy carriers. Hydrogen and other Power-to-X products [8] are considered as energy carrier.

Hydrogen has been recognized as a promising energy carrier due to its high energy density and potential for carbon-free energy conversion [9]. Establishing a hydrogen economy would involve the production of hydrogen by water splitting with electricity from renewable sources. Although hydrogen has a very high gravimetric energy density of 33.3 $kWh\ kg^{-1}$, its volumetric energy density is very low at 2.5 $Wh\ dm^{-3}$ at ambient pressure [10].

One of the major challenges associated with hydrogen is its storage and transportation [11]. This poses a significant barrier to the wider use of hydrogen as an energy carrier, especially for nonstationary applications. Liquid Organic Hydrogen Carriers (LOHC) have been proposed as hydrogen transport and storage medium. This requires reversible charging and discharging of hydrogen from suitable storage molecules. In almost all cases, appropriate catalytic systems are required to overcome kinetic barriers and accelerate chemical conversion.

In this context, ruthenium catalysts are discussed for their capacity to enable chemical hydrogen storage systems. In chemical hydrogen storage, hydrogen is stored in the form of a stable molecule and released on demand. Common examples are hydrocarbons, formic acid, and ammonia-borane, which can store hydrogen in a relatively dense, nongaseous form. Ruthenium catalysts have been shown to be effective in both the hydrogenation (for storage of hydrogen) and dehydrogenation (for release of hydrogen) of these compounds. In particular, the high activity of ruthenium-based catalysts can improve the energy efficiency of these processes, their high selectivity, the formation of unwanted by-products, and, overall, increase the hydrogen storage capacity.

2.1.2 Hydrocarbons as liquid organic hydrogen carriers

For hydrogen storage in hydrocarbons various storage molecules are being discussed at LOHC (**Table 1**). The LOHC has to fulfill a number of boundary conditions. A suitable LOHC should be capable of storing a large amount of hydrogen per unit volume and weight. The LOHC should be thermally and chemically stable in both its hydrogenated and dehydrogenated states. It should not degrade or react undesirably during the hydrogenation/dehydrogenation reaction or during storage and transportation. The LOHC system must be able to release the stored hydrogen on demand, and the spent carrier must be easily hydrogenated (**Table 2**). Therefore, the hydrogenation/dehydrogenation equilibrium should be in an appropriate temperature range, and the forward and backward reactions should be reasonably efficient. The LOHC should not be highly flammable, or toxic and should have a high flash point to ensure safe handling, storage, and transportation. Ideally, the LOHC should be environmentally benign, both in terms of its production and in its disposal or recycling. Lastly, the LOHC should be cost-effective to produce and recycle and ideally be derived from abundant or renewable sources.

Examples where ruthenium catalysts were explored for hydrogen storage include the use of benzyltoluene as LOHC (**Table 1**, Entry 1). The effect of the silica content in zirconia (ZrO_2) on the properties and hydrogenation activity of SiO_2-ZrO_2-supported ruthenium catalysts was investigated [14]. Catalysts were prepared with different Si/(Si + Zr) ratios from 0 to 30 mol%. Noteworthy, the textural properties, Ru particle size, and hydrogenation activity exhibit a volcano-shaped relationship with the Si content, with optimal results obtained at a ratio of 5 mol%. With a rising SiO_2 to ZrO_2 ratio, the thermal stability increased, and the particle size of tetragonal ZrO_2 decreased, which at a low ratio led to improved hydrogen storage efficiency. Further addition of SiO_2 led to decrease in the catalyst activity, along with increased surface heterogeneity and charge imbalance, which reduced the density of surface OH groups. Consequently, the addition of about 5 mol% SiO_2 to ZrO_2 appears to be optimal for enhancing the hydrogenation performance of Ru/ZrO_2 catalysts.

The selective hydrogenation of benzene to cyclohexene (**Table 1**, Entry 5) over supported ruthenium catalysts has been studied in a high-pressure slurry reactor [15], exploring suitable additives, supports, and reaction conditions to optimize cyclohexene yields. The addition of the organic base monoethanolamine (MEA) resulted in better selectivity than conventional inorganic salts such as zinc sulfate. Among different supports such as alumina, silica, titania, zirconia, and niobium oxide, alumina was found to be the best oxidic support. The presence of water was found to significantly improve the selectivity.

Entry	LOHC, parent form	m.p. [°C]	LOHC, reduced form	m.p. [°C]	n_{H2} [−]	w_{H2} [%]	H_2 storage [kWh dm^{-3}].	
1	Benzyl-toluene	−30	Perhydro-benzyltoluene	−70	3	3.21	1.66	[13]
2	*N*-Ethyl-carbazole	68.43	Perhydro-*N*-ethylcarbazole	<20	6	5.84	1.8	[13]
3	Toluene	−95.05 ± 0.6	Methyl-cyclohexane	−126.6 ± 0.4	3	6.16	1.59	[13]
4	Dibenzyl-toluene	−34	Perhydro-dibenzyltoluene	−45	9	6.24	1.94	[13]
5	Benzene	5.53 ± 0.08	Cyclohexane	6.45 ± 0.3	3	7.19	—	—
6	Naphthalene	80.1 ± 0.7	Decalin (trans/cis)	−31.2 −43.2	5	7.29	2.1	[13]

Table 1.
Selected liquid organic hydrogen carriers and their physical properties [12, 13].

Entry	Reaction	$\Delta H_R°$ [kJ mol^{-1}]	Eq.
1 (5)	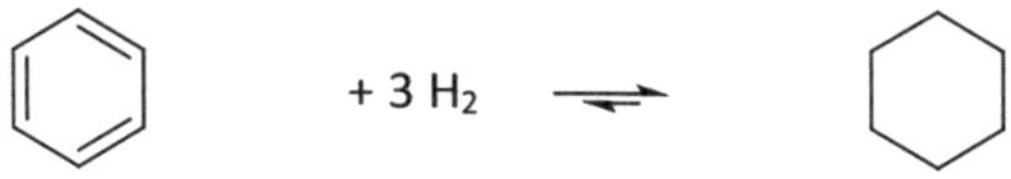	−205.3 ± 0.63 (6)	(1)
2	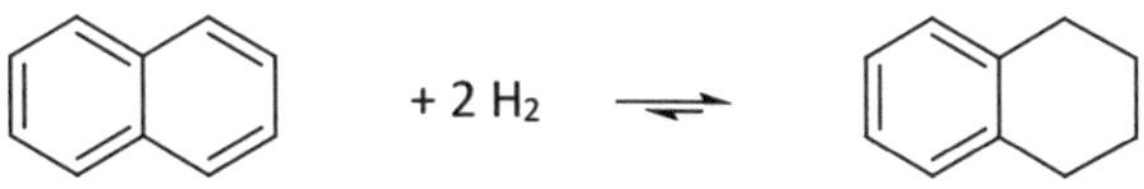	−120.5 ± 5.0 (6)	(2)
3	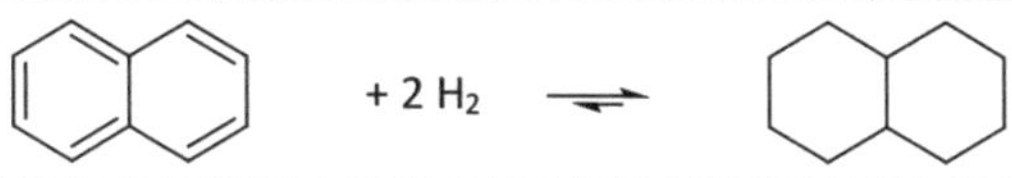	−318 (*cis* decalin) (6)	(3)
4 (2)	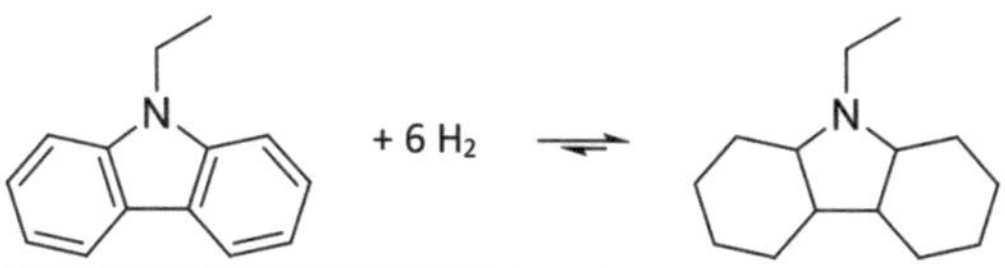	—	(4)

Table 2.
Reaction enthalpy for hydrogenation of selected LOHC. The entry number in brackets refers to ***Table 1****.*

2.1.3 Formic acid derivatives as liquid organic hydrogen carriers

Hydrogen can also be stored in formic acid by hydrogenating CO_2 to formic acid. CO_2-derived formic acid is a relevant candidate for hydrogen storage because it can carry 4.35 wt% hydrogen with excellent atomic efficiency [10]. The hydrogenation of CO_2 to formic acid is slightly exothermic and endergonic at standard conditions ($\Delta H_r° = -31{,}4$ kJ mol^{-1}, $\Delta G_r° = +31{,}8$ kJ mol^{-1}, [16]). This highlights the inherent thermodynamic instability of formic acid compared to the gaseous feedstocks. To overcome this thermodynamic hurdle, most reaction systems require the use of amines or other potent bases in stoichiometric amounts as co-reagents diminishing the weight percentage of hydrogen that is stored (**Figure 1**). The use of a stabilizing solvent such as dimethylsulfoxide (DMSO) was proposed as alternative [16].

Many molecular catalysts have been reported to highly efficiently convert CO_2 to formic acid, with turnover frequencies reaching several hundred thousand catalytic cycles per hour. The hydrogenation can be catalyzed by a number of ruthenium complexes, often in combination with specific ligands to improve performance. The

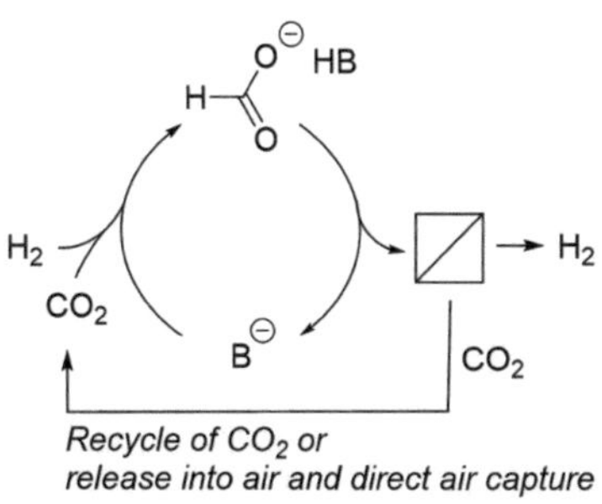

Base	Storage form	w_{H2} [%]
None	Formic acid	4.35
NH_3	$[HCO_2]^-$ $[NH_4]^+$	3.17
Lysine, K-salt	$[HCO_2]^-$ K^+·Lysine	1.15
Na_2CO_3	2 $[HCO_2]^-$ Na^+·H_2O	2.60

Figure 1.
Hydrogen storage cycle (left) based on formates as liquid organic hydrogen carrier (LOHC) and weight fraction w_{H2} of usable hydrogen in formic acid and various formate salts (right) [13, 17].

reverse process, the dehydrogenation of formic acid to release hydrogen and CO_2, can also be catalyzed by ruthenium. Importantly, the CO_2 released in this process can be captured and reused, creating a closed carbon cycle.

2.1.4 The ammonia-borane system as liquid organic hydrogen carriers

Ammonia-borane (NH_3BH_3) is another potential hydrogen storage option, with a remarkable theoretical hydrogen content of 19.6 wt% [18, 19]. Hydrolysis (Eq. (5)) or methanolysis (Eq. (6), [20–22]) are two practical methods of producing hydrogen from ammonia-borane for use in hydrogen fuel cells. In the presence of suitable catalysts, these reactions take place at ambient temperature under mild conditions. Ruthenium catalysts have been shown to be especially effective in promoting the dehydrogenation of ammonia-borane to release hydrogen. Despite these advantages, employing ammonia-borane as a hydrogen storage medium is not without its challenges. The primary issue lies in the regeneration of the spent fuel, which is often converted to a mixture of polymeric aminoborane and borazine. Regeneration requires substantial energy input and stringent conditions:

$$H_3NBH_3(aq) + 2\,H_2O(l) \rightarrow NH_4^+BO_2^-(aq) + 3\,H_2(g) \tag{5}$$

$$H_3NBH_3 + 4\,MeOH \rightarrow NH_4^+B(OMe)_4^- + 3\,H_2 \tag{6}$$

2.1.5 Role of ruthenium catalysts in hydrogen storage

Overall, the use of ruthenium catalysts in hydrogen storage can contribute to the development of sustainable hydrogen energy systems. By improving the efficiency and selectivity of chemical hydrogen storage processes, these catalysts can help to make hydrogen a more practical and sustainable energy carrier. However, challenges remain, including the development of more efficient catalysts, the improvement of fuel regeneration processes, and the integration of these systems with renewable energy sources and carbon capture technologies. With ongoing research in these areas, ruthenium catalysts will likely continue to play an important role in the future of hydrogen energy.

2.1.6 Hydrogenation of carbon dioxide

Carbon dioxide (CO_2) is an abundant, sustainable, and renewable feedstock [23] that holds great promise for the production of value-added products [24] spanning fuels, bulk and commodity chemicals, specialty products, and pharmaceuticals [25]. At the end of their life, when other recycling options are no longer feasible, carbon-based products are typically burnt to generate heat. Increased use of the released carbon dioxide as a feedstock could play a key role in closing anthropogenic carbon cycles [3].

CO_2, having a positively polarized carbon atom with two double bonds to electronegative oxygen atoms, is in the formal oxidation state of +IV. Therefore, except for the formation of carbonates [26] and carbonic acids [27], chemical transformations of CO_2 mostly involve the reduction of the carbon center [25]. The required reduction equivalents may be supplied by hydrogen or other reducing agents or may be provided electrochemically.

Considered a viable route for both CO_2 utilization and hydrogen storage, the hydrogenation of CO_2 to formic acid (Eq. (7)) has considerable potential [28]. Formic acid, a base chemical widely used in the chemical industry, can serve as a feedstock for fuel cells and as a medium for hydrogen storage. As of 2021, formic acid is a large-scale commodity with a production volume of 800 kt a^{-1} [29]. It has a theoretical hydrogen storage capacity of 53.4 g dm^{-3} [30]. A noteworthy aspect of using formic acid as an energy vector is that the CO_2 released during hydrogen production can be reused for formic acid production [30]. Thus, the hydrogenation of CO_2 to formic acid could contribute toward establishing a sustainable circular carbon economy.

The hydrogenation of carbon dioxide to formic acid (**Table 3**) is slightly exothermic but endergonic (Eq. (7)), [10, 31]). Also in the liquid phase, the reaction has a very low driving force (Eq. (8)). Consequently, basic co-reagents or solvents, such as ammonia or DMSO [16], respectively, are necessary to shift the chemical equilibrium toward the products either through salt formation (Eq. (9)) or stabilization of the formic acid. Ruthenium-based catalysts have shown great potential in the hydrogenation of CO_2 to formic acid, owing to their high activity and selectivity, coupled with high stability. Among them, supported ruthenium catalysts, especially those with ruthenium-hydroxide moieties, have shown superior performance. Nonetheless, the synthesis of these catalysts and the role of the hydroxyl groups in the catalytic reaction mechanism remain subjects of ongoing research.

For CO_2 hydrogenation with homogeneous catalysts, very high reaction rates have been reported including ruthenium [32, 33] and iridium [34] pincer complexes. In contrast, heterogeneous catalysts have recently received increased attention as many supported metal nanoparticles initially tested lacked substantial activity for the hydrogenation of CO_2 to formic acid.

A study of the synthesis of ruthenium-based catalysts for the hydrogenation of CO_2 to formic acid highlights the critical role of Ru-OH species in the reaction [22]. Detailed characterization of the catalysts using various spectroscopic and microscopic techniques revealed that hydroxyl groups strongly interact with the ruthenium moieties in the catalyst and influence its performance. Highly dispersed ruthenium-hydroxide species were beneficial for the CO_2 hydrogenation process, while the formation of crystalline RuO_2 species, which formed at high ruthenium content or high pH during catalyst preparation, hindered formic acid production. The study proposed an optimal set of conditions for catalyst preparation, including 2.0 wt% ruthenium loading, pH 12.8, and the use of aqueous ammonia as the stabilizing solvent.

For the efficient synthesis of liquid fuels (C5+ hydrocarbons) by CO_2 hydrogenation, homogeneous and heterogeneous catalysis have been combined [35] to overcome the challenges associated with heterogeneous catalysts for reverse water-gas shift (RWGS) and Fischer-Tropsch synthesis (FTS), which typically suffer from high operating temperatures and low selectivity. The system with combined homogeneous

Reaction	$\Delta H_R^0\left[\frac{kJ}{mol}\right]$	$\Delta G_R^0\left[\frac{kJ}{mol}\right]$	Eq.
$CO_2(g) + H_2(g) \rightleftharpoons HC(O)OH\ (l)$	−31.2	32.9	(7)
$CO_2(l) + H_2(l) \rightleftharpoons HC(O)OH\ (l)$	−31.2	−4	(8)
$CO_2(l) + H_2(l) + NH_3 \rightleftharpoons [HC(O)O^-][NH_4^+]\ (l)$	−84	−9.5	(9)

Table 3.
Enthalpy of reaction and change in free energy of CO_2 hydrogenation to formic acid derivatives. H2(l) refers to hydrogen dissolved in the aqueous solution.

RuCl3 and heterogeneous Ru0 catalysts achieved excellent C5+ selectivity (71.1%) at a low temperature of 180°C.

The publication reviews progress in the catalytic hydrogenation of CO_2 to C2+ hydrocarbons and oxygenates using both solid and molecular catalysts [36]. The study recognizes the value of these products due to their high energy density and compatibility with existing fuel infrastructure. However, the selective production of these compounds remains a challenge compared to C1 hydrogenation products such as methane and methanol. The work highlights the need for optimal catalytic functionalities integration for reductive and chain growth steps. The review also considers the thermodynamic and kinetic limitations of the process, in particular, the reverse water-gas shift reaction. It is suggested that future research should focus on metal carbide catalysts and liquid phase molecular catalysts to decouple the chain growth kinetics from the gas phase CO concentration.

Merging the fields of homogeneous and heterogeneous catalysis could be a way to improve the efficiency of CO_2 hydrogenation. These two fields of catalysis have developed separately, although they share common challenges and mechanisms. A "catalytic ensemble" of a dual ruthenium catalyst, in which a homogeneous and a heterogeneous catalyst work in tandem, works well for the conversion of CO_2 to hydrocarbons of different chain lengths [37]. A high hydrocarbon yield of more than 70% is obtained under mild conditions. The key questions concern the mechanism of this improved performance, the role of lithium in the reaction, and the feasibility of replacing rare metals with sustainable alternatives.

The performance of two ruthenium-based catalysts, Ru/SiO_2 and Ru/MCM-41, for the selective hydrogenation of CO_2 to formic acid, is compared [38]. The researchers used well-established protocols for the synthesis of these catalysts and analyzed their physicochemical properties using a variety of analytical techniques. The Ru/MCM-41 system was found to be more efficient than the Ru/SiO_2 system in terms of turnover number (TON) and turnover frequency (TOF). Functional ionic liquids were used as both reaction medium and absorbent for CO_2 solubilization and formic acid anchoring. This allowed the reaction to proceed in a more optimized manner, improving the selectivity and recyclability of the catalyst. The ionic liquid di(N,N-dimethylaminoethyl)-2-methylimidazolium (DAMI) sulfonate [DAMI][$CF_3(CF_2)_3SO_3$] proved to be particularly effective due to its high CO_2 absorption capacity. The use of diamine-functionalized ionic liquids also improved the coordination with formic acid. The Ru/MCM-41 catalyst in the ionic liquid [DAMI][$CF_3(CF_2)_3SO_3$] provided the highest TON of formic acid.

An industrially viable approach to the synthesis of *N,N*-dimethylformamide (DMF) *via* CO_2 hydrogenation involves the use of a ruthenium-grafted bisphosphine-based porous organic polymer (Ru/PP-POP) catalyst [39]. Historically, CO_2 conversion methods have been difficult to scale up for industrial use. The grafted Ru catalyst was shown to be highly efficient and recyclable. The catalyst developed in this work exhibits remarkable activity, achieving an unprecedented turnover number of 160,000 and an initial turnover frequency of 29,000 h^{-1} in a batch process. Furthermore, the catalyst shows excellent stability and selectivity in both batch and continuous flow processes. The continuous flow process was carried out in a trickle bed reactor which achieved a high productivity of 915 mmol g_{Ru}^{-1} h^{-1}. The breakthrough presented in this research may pave the way for the large-scale production of DMF, a commonly used industrial solvent, directly from CO_2.

The hydrogenation of CO_2 can be performed in a variety of solvents, including CO_2-rich scrubbing solvents from carbon capture units. Such solvents are capable of capturing CO_2 from flue gas streams, making this process potentially useful for carbon capture and utilization (CCU) strategies. However, the direct hydrogenation of CO_2-rich scrubbing solvents remains challenging due to other components in the gas stream that can act as catalyst poisons. Furthermore, the hydrogenation of CO_2 to formic acid is reversible, which limits the net production of formic acid.

The research focuses on the synthesis and evaluation of alumina-supported Ru nanoparticles as catalysts for the selective hydrogenation of CO_2 to formic acid [40]. The Ru/Al_2O_3 (2–10 wt% Ru) catalysts were prepared by the ethylene glycol reduction method, and their physicochemical properties were thoroughly investigated using techniques such as XRD, transmission electron microscopy (TEM), EDX, H_2 chemisorption, XPS and H_2-TPD analysis. The catalysts showed promising activity and selectivity for CO_2 hydrogenation, with the performance closely related to their physicochemical properties, especially the Ru metal dispersion. In particular, the Ru/Al_2O_3 catalyst outperformed others when used in conjunction with the ionic liquid [DAMI] [$CF_3(CF_2)_4SO_3$]. This medium not only increased catalyst activity but also acted as a stabilizer, preventing catalyst deactivation and promoting formic acid formation via intermediate carbonate species. The multifunctional ionic liquid also facilitated the recovery and recycling of the heterogeneous catalyst. The Ru/Al_2O_3 catalyst showed remarkable stability and could be recycled up to eight times without significant catalyst leaching. This study therefore provides an effective, recyclable catalytic system for the conversion of CO_2 to formic acid, which has potential implications for CO_2 utilization and sustainable chemistry.

The synthesis and successful implementation of a heterogenized molecular Ru catalyst on bpyTN-30-CTF support the continuous hydrogenation of CO_2 to formic acid [10]. The support provides increased porosity and metal anchoring sites, enabling remarkable catalytic performance for commercialization. In an integrated trickle bed reactor system, the catalyst achieved a high productivity of 669.0 g ${g_{cat}}^{-1}$ d^{-1} and a CO_2 conversion of 44.8%. Notably, the catalyst showed excellent stability over 30 days of operation, reaching a total turnover number of 524,000 without significant deactivation. The impressive performance and stability of the catalyst in a fixed-bed multiphase reactor system demonstrate the high industrial feasibility of this approach. The integrated pilot scale system for the production of pure formic acid, including a separation unit for the formate adduct, could provide a platform for an industrially viable CO_2 hydrogenation process.

2.2 Ruthenium catalysts in amine synthesis

A ruthenium-tungsten bimetallic catalyst was introduced for the hydrogenation of aliphatic, acyclic primary amides to primary amines, a key transformation for potential bio-based amine production [41]. The partial pressure of ammonia was identified as a key parameter in achieving high yields of primary amines, with up to 83% yield of hexylamine from hexanamide. The study examines the effects of catalyst support, platinum group metal to Lewis-acid ratio, hydrogen pressure, temperature, solvent tolerance, and product stability. The results show the ruthenium-tungsten catalyst is suitable for hydrogenating a wide range of primary amides.

2.3 Hydroformylation reactions

Hydroformylation, also known as the oxo process, is an important industrial process for the synthesis of aldehydes from alkenes, carbon monoxide, and hydrogen.

The reaction is typically catalyzed by transition metals, traditionally cobalt-based catalysts or, more recently, rhodium-based catalysts. However, ruthenium-based catalysts have also been investigated for this reaction. Ruthenium catalysts have some specific advantages in hydroformylation. They generally exhibit high activity and selectivity and are often resistant to sulfur and other impurities. This can make them particularly suitable for hydroformylation reactions involving difficult substrates or challenging reaction conditions. In addition, some ruthenium catalysts have been found to promote the isomerization of alkenes prior to hydroformylation, resulting in branched aldehydes, which may be desirable in certain applications. Despite these advantages, the use of ruthenium catalysts in industrial hydroformylation processes is less common than that of cobalt or rhodium catalysts. This is primarily due to the higher cost associated with ruthenium and its less explored nature in this specific application.

2.4 Metathesis reactions

Ruthenium catalysts have emerged as key agents in olefin metathesis, a widely used reaction in synthetic chemistry in which alkylidene groups are exchanged between different olefins to create a wide variety of complex molecular structures. Two examples shall be given below:

A highly active and reusable heterogeneous ruthenium catalyst specifically designed for olefin metathesis reactions involves ruthenium complexes [$L_2X_2Ru = CHR$], where L represents various ligands including triphenylphosphine, tricyclohexylphosphine, or *N*-heterocyclic carbenes, and X is a chlorine atom, all grafted onto the mesoporous silica material SBA-15 [42]. The unique approach is to anchor these ruthenium complexes within the pore channels of the SBA-15 material. This arrangement effectively prevents the decomposition of the catalytic species, a common problem with such catalysts. The result is a catalyst that not only exhibits impressive catalytic activity in olefin metathesis reactions but also shows significant durability. The activity of the catalyst does not decrease even after repeated use.

A promising avenue for the preparation of effective ruthenium-based metathesis catalysts is homogeneous and heterogeneous ruthenium-based metathesis catalysts that feature electron-withdrawing ligands. Such Grubbs–Hoveyda type metathesis catalysts can be prepared by chlorine exchange, replacing one or two chlorine ligands with trifluoroacetate and trifluoromethanesulfonate anions [43]. These catalysts exhibit higher reactivity in ring-closing metathesis (RCM) at 45°C compared to previous ruthenium-based catalysts, achieving turnover numbers (TONs) up to 1800 in RCM. They also showed superior performance in enyne metathesis and ring-opening cross-metathesis with norborn-5-ene and 7-oxanorborn-5-ene-derivatives. Notably, one of the catalysts exhibited significant RCM activity even at room temperature. In addition, heterogeneous catalysts were prepared by immobilization on polystyrene-divinylbenzene support. Especially, a variant with both chlorine ligands replaced showed high activity. Importantly, the leaching of ruthenium into the reaction mixture was exceptionally low, resulting in virtually Ru-free products.

Ruthenium-catalyzed olefin metathesis could play an important role in sustainable chemistry and the energy transition. For example, ruthenium metathesis catalysts can be used in reactions involving renewable feedstocks such as plant-derived oils and fats. This enables the production of bio-based chemicals and fuels, contributing to the transition to renewable energy sources. Olefin metathesis can also play a role in recycling recovered gases such as lower olefins into valuable chemicals, creating carbon-neutral cycles.

2.5 Use of ruthenium catalysts in oxidation reactions

Due to their unique reactivity profiles, ruthenium catalysts can play a versatile role in oxidation reactions, enabling diverse applications in sustainable chemical synthesis. In particular, the oxidized form RuO_2 has a wide range of potential applications in oxidation reactions, ranging from zero energy air purification, low-temperature fuel cells, electrochemical water splitting, to the oxidative dehydrogenation of simple alcohols [44]. The low-temperature oxidation of hydrochloric acid (HCl) to chlorine and of ammonia (NH_3) to nitrogen monoxide (NO) is also catalyzed. Oxidation of carbon monoxide over supported Ru and single crystal ultrathin RuO_2 films [44] revealed that the most active and stable state is an ultrathin RuO_2 shell coating with a metallic Ru core. Catalytic conversions tend to cause structural deactivation of Ru-based catalysts.

There are different strategies for immobilizing redox-active elements in solid matrices for application in oxidation reactions including framework substitution, grafting or sol–gel methods, and ion exchange in layered double hydroxides [45]. In oxidations with O_2, H_2O_2, and RO_2H as primary oxidants, ruthenium catalysts are a good choice among other metal catalysts such as titanium, chromium, cobalt, manganese, iron, tungsten, molybdenum, vanadium, and tantalum. Many of these catalyst systems, particularly those involving oxometal species, are prone to leaching. Elements present as oxometal species are more susceptible to leaching due to their less stable coordination environments. Variable valence metals appear to be more resistant to solvolysis, but this is only true for low-conversion oxidations. In addition, secondary oxidation products can cause leaching of metal ions. Consequently, the heterogeneous nature is often questioned, and the search for heterogeneous catalysts with unique activities, selectivities, and operational stability in liquid phase oxidations is ongoing.

Ruthenium oxide-based catalysts supported on activated carbon (AC) and zeolite ZSM-5 were shown to be active for the degradation of aqueous phenol. RuO_2/AC showed superior performance to $RuO_2/ZSM5$ [46]. Complete phenol degradation and 60% total organic carbon (TOC) removal were achieved within one hour. The catalysts activate peroxymonosulfate and effectively generate sulfate radicals that are responsible for phenol degradation. Phenol degradation followed a pseudo-first-order kinetics with activation energies of 61.4 kJ mol^{-1} for RuO_2/AC and 42.2 kJ mol^{-1} for $RuO_2/ZSM5$.

Ruthenium cations, in combination with microcrystals of cobalt hydroxide and cerium oxide, catalyze the oxidation of various types of alcohols to carbonyl compounds. Using atmospheric oxygen, the reaction occurs under mild conditions at atmospheric pressure and 60°C [47]. Notable is the efficient conversion of primary aliphatic alcohols to the corresponding carboxylic acids in high yields.

The selective aerobic oxidation of ethanol to acetic acid is of considerable industrial importance, since acetic acid is a fundamental bulk chemical and food ingredient. The system is also interesting for bioethanol upgrading. Ruthenium-hydroxide on ceria, 1.2 wt % $Ru(OH)_x/CeO_2$, provided acetic acid in quantitative yields under optimal conditions (150°C, 10 bar O_2, 12 h reaction time, molar ratio Ru/substrate 0.23) [48]. Ceria was found to be the most effective support on different support materials (titania, alumina, ceria, and spinel).

Ru/Al_2O_3 is a good catalyst for the oxidation of alcohols to carbonyl compounds [49]. The catalyst uses molecular oxygen as the oxidant and tolerates both activated and nonactivated alcohols, even those containing sulfur or nitrogen atoms or a carbon-carbon double bond. Oxidation of alcohols under atmospheric conditions has been achieved without the need for additives.

Water-tolerant catalysts are required to overcome the current limitations for practical lignin valorization. In this context, ruthenium supported on γ-alumina or silica has been shown to be an effective catalyst for the oxidation of veratryl alcohol, a compound formed by cleavage of β-O-4 linkages [4] in lignin, to veratraldehyde [50]. The Ru/Al_2O_3 catalyst, prepared with ruthenium(IV)oxide hydrate, showed superior catalytic activity, yielding 89% yield in veratraldehyde in water at 160°C under 5 bar air pressure after 8 hours. Longer reaction times resulted in significant decarbonylation of veratraldehyde to veratrol. The use of methanol as a solvent instead of water prevented the oxidation of the hydroxyl group in veratraldehyde, indicating that methanol has a protective effect. Catalysts with other transition metals (Mn, Co, Cu, and Ag) showed significantly lower activities compared to Ru/Al_2O_3.

2.6 Synthesis of fine chemicals

Due to their exceptional catalytic activity, selectivity, and stability, ruthenium catalysts are poised to improve the synthesis of fine chemicals, increasing efficiency and sustainability in this sector of the chemical industry.

The use of ruthenium catalysts shall be exemplified in two distinct contexts for the synthesis of fine chemicals, the hydrogenation of α-amino acids and the epoxidation of olefins using hydrogen peroxide as a 'green' reoxidant [51]. A bimetallic Ru/Re sponge catalyst was used for the stereoretentive hydrogenation of α-amino acids in aqueous medium. The bimetallic catalyst was supported on carbon. Efficient synthesis of amino alcohols with high enantiomeric excess was achieved. Relatively low temperatures prevented racemization. For epoxidation, ruthenium pyridine-2,6-dicarboxylate complexes with tridentate *N*-donor ligands served as homogeneous catalysts and provided protocols for both asymmetric and nonasymmetric reactions. This study underscores the principles of green chemistry, emphasizing the atom efficiency of this conversion and the use of water as a solvent.

Ruthenium catalysis often has distinct capabilities over other catalytic systems such as palladium, rhodium, iridium, or cobalt complexes. Meta-selective remote C – H bromination on aryl-substituted purines has been catalyzed using the heterogeneous ruthenium Ru/SiO_2 catalyst [52]. This exemplifies the utility of ruthenium-catalyzed C – H activation and paves the way for novel strategies in direct nucleobase fluorescent labeling of purines, a relevant area in biochemistry and molecular biology. It also highlights the synthetic advantages of ruthenium-catalyzed C – H activation and illustrates the unique potential of ruthenium catalysts in the synthesis of fine chemicals.

A recent study examined the use of ruthenium catalysts for the Meerwein–Ponndorf–Verley reaction and the isomerization of allylic alcohols to saturated ketones. In this study, a ruthenium metal-organic framework (Ru/MOF) catalyst was modified through defect engineering and hydrogen pretreatment [53]. This enhancement was achieved by using a mixed-linker approach on the Ru-MOF $[Ru_3(BTC)_2A_n]\cdot G_m$ (where BTC represents benzene-1,3,5-tricarboxylate; A is a counterion; G stands for a guest molecule). The mixed-linker approach introduced structural defects at the Ru paddlewheel nodes, resulting in the formation of partially reduced Ru-nodes and enhanced catalytic activity. Hydrogen pretreatment further enhanced the catalytic activity.

Several features make ruthenium catalysts of particular interest for the synthesis of fine chemicals: The versatility of ruthenium catalysts makes them useful for the synthesis of many different types of fine chemicals. Ruthenium catalysts can often provide high levels of chemo-, regio-, and stereoselectivity, enabling the production of fine chemicals with precise structural features. Ruthenium catalysts are generally tolerant of a wide range of functional groups and reaction conditions, allowing for broad substrate scope and operational simplicity. Together with the general aspects discussed above, these factors make ruthenium catalysts an attractive option for the synthesis of fine chemicals, contributing to greener and more sustainable chemical processes.

3. Interaction of ruthenium particles with the support

A study of the nature of the catalyst challenged previous claims of homogeneous benzene hydrogenation with Ru(arene) precatalysts [54]. The study provided evidence that the real catalyst in benzene hydrogenation is bulk ruthenium metal particles, not a homogeneous metal complex or a soluble nanocluster. The reaction was initiated with $Ru(II)(\eta^6\text{-}C_6Me_6)(OAc)_2$ as the precatalyst. Evidence that the catalysis is based on nanoparticles is provided by (i) the nucleation and autocatalytic surface-growth mechanism; (ii) the formation of bulk ruthenium metal during hydrogenation; (iii) the high activity of bulk ruthenium metal; (iv) inactivity of the filtrate until bulk metal is formed; (v) complete inhibition of catalysis by Hg(0), a known poison for heterogeneous catalysts; and (vi) absence of detectable nanoclusters under conditions where they are typically found. The study suggests that a similar investigation be conducted for other benzene hydrogenation catalysts derived from Ru(arene) precatalysts.

In accordance, supported ruthenium nanoparticles are used as heterogeneous catalysts. Metal nanoparticles must always be stabilized against agglomeration [55]. To stabilize the ruthenium nanoparticles in a supported ruthenium nanoparticle catalyst, an intimate interfacial contact between the Ru nanoparticles and the support is required for stable performance. Such interactions can be induced by appropriate choice and modification of the oxide or carbon support. A suitable method for preparing supported ruthenium catalysts usually involves special thermal reduction protocols.

A study of the selective oxidation of 5-hydroxymethylfurfural (HMF) to 2,5-furandicarboxylic acid underscores the importance of support materials in the design of efficient catalysts for the sustainable conversion of biomass-derived compounds into valuable chemicals. HMF is a biomass-derived chemical typically produced by the dehydration of sugars and polysaccharides. Heterogeneous ruthenium-based catalysts have been shown to be very effective for the oxidation of HMF with molecular oxygen in the absence of a base and in water as a solvent [56]. Among a wide range of support materials, including various metal oxides, magnetite, spinel, hydrotalcite, and hydroxyapatite, ruthenium-hydroxide species supported on cerium oxide exhibited superior catalytic activity and selectivity compared to those using titanium dioxide as support.

In the case of carbon supports, Ru precursors were deposited on carbon materials and then reduced to Ru nanoparticles, resulting in catalysts with exceptional performance in the hydrogenation of benzene and toluene [57]. Remarkably, these thermally treated Ru catalysts operate up to 24 times faster than their conventionally prepared counterparts and exhibit impressive resistance to oxidation, reduced leaching, and well-distributed dispersion on carbon materials.

A magnetically separable ruthenium-hydroxide catalyst on magnetite ($Ru(OH)_x/Fe_3O_4$) has been used for three types of reactions: aerobic oxidation of alcohols, aerobic oxidation of amines, and reduction of carbonyl compounds to alcohols using 2-propanol as a hydrogen donor [58]. This shows that the catalyst is effective for a wide range of substrates, including aromatic, aliphatic, and heterocyclic substrates. A significant advantage of this catalyst is its magnetic separability: the $Ru(OH)_x/Fe_3O_4$ catalyst can be easily separated from the reaction products using a permanent magnet, with a recovery of more than 99%. The study confirmed the intrinsic heterogeneity of the catalysis for these reactions and showed that the $Ru(OH)_x/Fe_3O_4$ catalyst retained its activity after multiple reuses.

For binding ionic ruthenium complexes [59] and ruthenium clusters, nitrogen-doped carbon materials have attracted attention due to their strong interaction with the active metal species through their nitrogen functionalities. Such nitrogen-doped carbon materials can be prepared from adenine as carbon and nitrogen source and anhydrous magnesium chloride as template [59]. Favorable binding sites include pyridinic and pyrrolic nitrogen moieties (**Figure 2**).

As a biorenewable, biodegradable material with multiple functional groups, chitosan is an attractive candidate for anchoring metal catalysts. Ruthenium supported on chitosan, a natural polysaccharide, has been used as a heterogeneous catalyst for the hydration of nitriles to amides in aqueous media under neutral conditions [60]. The chitosan-ruthenium catalyst offers high yields, excellent selectivity, and easy recoverability, all under neutral, environmentally benign conditions.

Ruthenium macromolecules, such as ruthenium tetraphenylporphyrin [Ru(TPP)], have been shown to be an effective catalyst for the reaction of carbon dioxide (CO_2) with epoxide to form cyclic carbonates [61]. The [Ru(TPP)] catalyst showed excellent performance, following first-order kinetics with respect to epoxide, leading exclusively to cyclic carbonate. Importantly, it offered significantly lower activation energy than previously reported catalysts. For heterogenization, the [Ru(TPP)] complex was grafted onto functionalized SBA-15 molecular sieves. When grafted onto aminosilane-

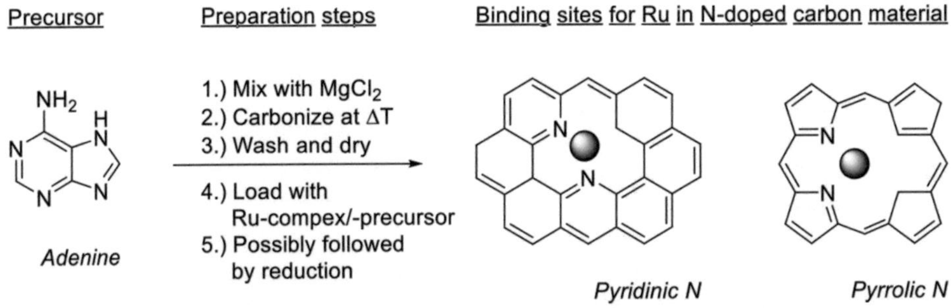

Figure 2.
Synthesis of N-*doped carbon materials and binding sites according to [59].*

and iodosilane-functionalized materials, [Ru(TPP)] showed promising catalytic activity for the carboxylation of epoxides.

Heterogeneous ruthenium catalysts have a significant potential for efficient and selective chemical transformations with relevance to environmental sustainability. A heterogeneous ruthenium catalyst prepared by adsorption of $[RuCl_2(p\text{-cymene})]_2$ on activated carbon showed exceptional efficiency and selectivity for the aerobic oxidation of alcohols, hydrolytic oxidation of silanes, and the dehydration of aldoximes [62].

The use of agricultural wastes in the preparation of carbon supports has also been explored as a way to increase the sustainability of catalyst production. One example is a heterogeneous ruthenium catalyst using silica derived from rice husk ash (RHA) [63]. The ruthenium was loaded by precipitation using a metal salt solution in nitric acid followed by high-temperature calcination. The structure varied depending on the calcination protocol. Initial analysis showed that the Ru/RHA catalyst was in an amorphous state. After postcalcination at 700°C, some degree of crystallization was observed. SEM images revealed the formation of nano-sized rods within the amorphous powder. Interestingly, the specific surface area decreased significantly upon postcalcination, consistent with the formation of the rod-shaped crystalline phase. The catalyst exhibited strong metal-oxygen chemical bonding, which is believed to be the key interaction between the metal and the silica support. The inhomogeneous Ru distribution in the initial Ru/RHA catalyst improved upon calcination, as shown by EDX.

The study thus showcases the successful development of a series of heterogeneous ruthenium catalysts that possess not only comparable activity to homogeneous ones but also offer practical advantages like easy separation and reuse. These catalysts contribute to flexible hydrogen generation and hold the potential for selective and efficient formic acid decomposition in diluted solutions.

An interesting heterogeneous ruthenium catalyst for the selective decomposition of formic acid into hydrogen and carbon dioxide comprised a combination of a Ru(II) *meta*-trisulfonated triphenylphosphine (*m*TPPTS) complex with phosphine-modified mesoporous silica [64]. Variations in the catalyst structure were achieved by changing the length of the alkylene chain connecting the silica to the diphenylphosphine group. The catalysts were highly active, with Ru-*m*TPPTS/MCM41-$Si(CH_2)_2PPh_2$ exhibiting a turnover frequency of 2780 h^{-1} at 110°C, comparable to the corresponding homogeneous catalyst. No evidence of ruthenium leaching was detected even after 71,000 turnovers. This shows remarkable durability.

4. Bimetallic catalysts and catalyst mixtures

Alloying ruthenium with other elements imparts unique properties and enhanced reactivity profiles to bimetallic ruthenium catalysts. The integration of additional metals imparts specific properties to ruthenium that profoundly alter reaction pathways compared to their monometallic counterparts.

A review on the synthesis and analysis of polynuclear ruthenium-tin cluster complexes focuses on their transformation into heterometallic nanoparticles [65]. These complexes are then used as catalysts in the hydrogenation of unsaturated organic molecules. The narrative of the study bridges the gap between the molecular design of these clusters and their use in facilitating industrially relevant hydrogenation reactions, reflecting the importance of this class of bimetallic nanoscale heterogeneous catalysts. Unique properties are imparted to the catalyst by the cooperative action of

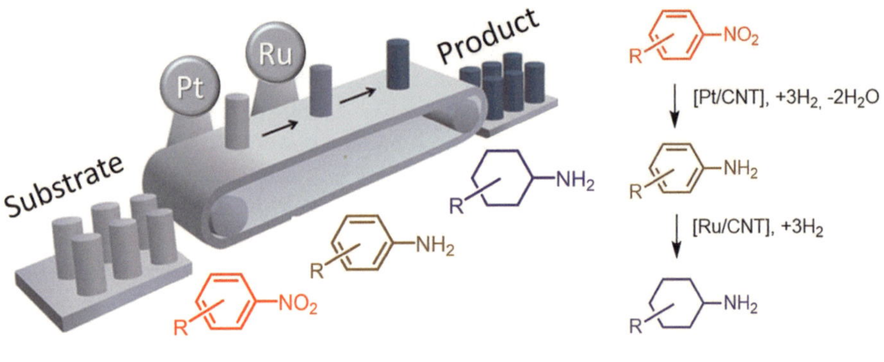

Figure 3.
Concept of a molecular assembly line of a physical mixture of two different heterogeneous catalysts applied to the hydrogenation of nitrobenzene over a Pt/CNT Ru/CNT catalysts mixture.

ruthenium and tin, providing insight into how these properties might be exploited for more efficient and selective hydrogenation processes.

Uniformly distributed bimetallic Ni-Ru nanoparticles on a magnesium-aluminum hydrotalcite support are effective in reducing nitrobenzene under ambient conditions [66], demonstrating that the presence of metallic nickel and ruthenium on the hydrotalcite surface facilitates this process. Kinetic studies indicate a first-order relationship with respect to nitrobenzene. Importantly, the catalyst maintains its activity over multiple cycles and effectively reduces various substituted nitroarene molecules.

Likewise, physical mixtures of different heterogeneous catalysts can alter the course of the reaction. Thus, carbon nanotube-supported Ru/CNT and Pt/CNT catalysts have been effectively employed in the hydrogenation of nitroaromatics to cycloaliphatic amines [67]. Interestingly, Ru and Pt, as metals, share similar atomic radii of 133 and 137 pm, while their static average electric dipole polarizabilities differ [68]. Consequently, the aromatic ring, categorized as "soft" due to its aromatic π-system dispersed over six carbon atoms, tends to adsorb preferentially on the more polarizable ruthenium. The nitro group, deemed "hard" due to its negative charge spread mostly over only two oxygen atoms, tends to adsorb preferentially on platinum, which has highly shielded d-electrons. A combination of 95% Ru/CNT and 5% Pt/CNT provides approximately equal rates for the hydrogenation of the respective moieties and maximum selectivity toward the fully hydrogenated product, cyclohexylamine (**Figure 3**).

Bimetallic ruthenium catalysts, thus, often offer distinct advantages including enhanced catalytic performance and an increased range of reactivity.

5. Conclusions

Ruthenium catalysts have demonstrated a remarkable capacity for a multitude of reactions, encompassing hydrogenation, oxidation processes, and, notably, the contemporary utilization of carbon dioxide. The ability of ruthenium catalysts to convert carbon dioxide into useful chemicals and fuels may not only help to sequester CO_2, a potent greenhouse gas but also produce valuable feedstocks for industry, providing a potential route to carbon-neutral or even carbon-negative processes. Similarly, ruthenium catalysts may play an increased role in the intensified use of bio-derived feedstock. In the field of electrocatalysis, ruthenium shows particular promise in water-splitting technologies, organic electrosynthesis, proton exchange membrane fuel cells, and metal-air batteries. The metal's propensity to catalyze the oxygen evolution reaction significantly increases the efficiency of hydrogen production.

The interplay between homogeneous and heterogeneous catalysis using ruthenium, as demonstrated in CO_2 conversion, also illustrates the immense potential of the metal to unlock new technologies and sustainable solutions. It is the ability to form bonds of intermediate strength with a variety of ligands, exist in multiple oxidation states, and accommodate both high and low oxidation states that enables this wide range of chemical transformations. Delving deeper into the fundamental atomic-level mechanisms could pave the way for further advances in the design of ruthenium catalysts, thereby expanding their efficacy and applicability. Particularly in the global raw material transition to sustainable feedstocks and the energy transition to more sustainable and cleaner energy systems, ruthenium catalysts offer a range of solutions in key areas that can make a significant contribution to reducing carbon emissions and optimizing energy use.

In summary, the extensive exploration of ruthenium as a catalyst underscores its fundamental role in modern chemistry and materials science. While current advances in ruthenium catalysis are impressive, the breadth of its untapped potential encourages continued exploration and innovation. As researchers develop an even more refined understanding of its properties, ruthenium is likely to continue to be at the forefront of new discoveries and applications, contributing to sustainable solutions in exciting areas of technology.

Acknowledgements

This contribution was made possible by funding of the state of North Rhine-Westphalia, grant number IRR-2018-1, RWE Power AG, and the Faculty of Mechanical Engineering of Ruhr-Universität Bochum as part of the endowed chair CSC. Also the financial support of the Bundesministerium für Bildung und Forschung (BMBF) for the project Production of Ethene from Recycling Streams and Renewable Carbon Sources as a Sustainable and Economic Route to Basic Building Materials for the Chemical Industry (Syngas2Ethene, grant number 01LJ2107A) is acknowledged. TEM would like to thank Prof. M. Dröscher for the stimulating scientific discussions on the subjects of raw material and energy supply to the chemical industry.

Conflict of interest

The author declares no conflict of interest.

Acronyms and abbreviations

AC	Activated Carbon
CCU	Carbon Capture and Utilization
PP	Bisphosphine
EDX	Energy Dispersive X-Ray Spectroscopy
ΔH_r°	Heat of reaction at standard conditions
ΔG_r°	Gibbs free energy at standard conditions
L	Ligand (L)

LOHC	Liquid Organic Hydrogen Carriers
OER	Oxygen Evolution Reaction
ORR	Oxygen Reduction Reaction
PEMFC	Proton Exchange Membrane Fuel Cells
m.p.	Melting Point
MOF	Metal-Organic Framework
n_{H2}	Mols of Hydrogen
MEA	Monoethanolamine
POP	Porous Organic Polymer
RHA	Rice Husk Ash
RCM	Ring-Closing Metathesis
TPD	Temperature Programmed Desorption
TOC	Total Organic Carbon
TEM	Transmission Electron Microscopy
TOF	Turnover Frequency
TON	Turnover Number
Wh	Watthours
w_{H2}	Weight Percentage of Hydrogen
XRD	X-Ray Diffraction
XPS	X-ray Photoelectron Spectroscopy
OAc_-	Acetate
ZSM-5	Aluminosilicate Zeolite Socony Mobil–5
BTC	Benzene-1,3,5-tricarboxylate
DMF	*N,N*-Dimethylformamide
DMSO	Dimethylsulfoxide
DAMI	Di(*N,N*-dimethylaminoethyl)-2-methylimidazolium
HMF	5-Hydroxymethylfurfural
MCM-41	Mesoporous Material Mobil Composition of Matter No. 41
TPP	Tetraphenylporphyrin
*m*TPPTS	*meta*-Trisulfonated triphenylphosphine

Author details

Thomas Ernst Müller
Ruhr-Universität Bochum, Bochum, Germany

*Address all correspondence to: mueller@ls-csc.rub.de

References

[1] Ogba OM, Warner NC, O'Leary DJ, Grubbs RH. Recent advances in ruthenium-based olefin metathesis. Chemical Society Reviews. 2018;**47**(12): 4510-4544

[2] Chelucci G. Ruthenium and osmium complexes in CC bond-forming reactions by borrowing hydrogen catalysis. Coordination Chemistry Reviews. 2017; **331**:1-36

[3] Tomkins P, Müller TE. Evaluating the carbon inventory, carbon fluxes and carbon cycles for a long-term sustainable world. Green Chemistry. 2019;**21**(15): 3994-4013

[4] Panke D, Bechthold G, Müller TE. Solvent effect in catalytic lignin Hydrogenolysis. Catalysts. 2022;**12**(6): 664-681

[5] Syngas2Ethene, Closing the Carbon Cycle: Production of Ethene from Recycle Streams and Renewable Carbon Sources as a Sustainable and Economic Route to Feedstock for the Chemical Industry. Available from: https://www.fona.de/de/massnahmen/foerdermassnahmen/KlimPro/SynGas2Ethen.php, https://reinvent-klimpro.de/KLIMPRO_Projekte/SynGas2Ethen.html, https://www.heraeus.com/en/group/innovation/press_and_news_1/2022_3/green_chemistry.html

[6] Nature Human Behaviour. Climate change and human behaviour. 2022;**6**: 1441-1442

[7] Dreyfus GB, Xu Y, Shindell DT, Ramanathan V. Mitigating climate disruption in time: A self-consistent approach for avoiding both near-term and long-term global warming. PNAS. 2022;**119**(22): e2123536119

[8] Hermesmann M, Grübel K, Scherotzki L, Müller TE. Promising pathways: The geographic and energetic potential of Power-to-X technologies based on regeneratively obtained hydrogen. Renewable and Sustainable Energy Reviews. 2021;**138**:110644

[9] Hermesmann M, Müller TE. Green, turquoise, blue, or grey? Environmentally friendly hydrogen production in transforming energy systems. Progress in Energy and Combustion Science. 2022;**90**:100996

[10] Park K, Gunasekar GH, Kim S-H, Park H, Kim S, Park K, et al. CO_2 hydrogenation to formic acid over heterogenized ruthenium catalysts using a fixed bed reactor with separation units. Green Chemistry. 2020;**22**(5):1639-1649

[11] Hermesmann M, Tsiklios C, Müller TE. Environmental assessment of climate-friendly hydrogen supply chains–a trade-off between capacity utilization and transport distance? Energy. 2022;**2004**:2965

[12] Lemmon EW, Bell IH, Huber ML, McLinden MO. WebBook NC. NIST Standard Reference Database Number 69. 2023. DOI: 10.18434/T4D303

[13] Grubel K, Su J, Kothandaraman J, Brooks K, Somorjai GA, Autrey T. Research requirements to move the Bar forward using aqueous Formate salts as H2 carriers for energy storage applications. Journal of Energy and Power Technology. 2020;**02**(04):016

[14] Kim TW, Kim C, Jeong H, Shin C-H, Suh Y-W. Hydrogen storage into monobenzyltoluene over Ru catalyst supported on SiO2-ZrO2 mixed oxides with different Si/Zr ratios. Korean

Journal of Chemical Engineering. 2020; **37**(8):1427-1435

[15] Suryawanshi PT, Mahajani VV. Liquid-phase hydrogenation of benzene to cyclohexene using ruthenium-based heterogeneous catalyst. Journal of Chemical Technology and Biotechnology. 1997;**69**(2):154-160

[16] Rohmann K, Kothe J, Haenel MW, Englert U, Hölscher M, Leitner W. Hydrogenation of CO_2 to formic acid with a highly active ruthenium Acriphos complex in DMSO and DMSO/water. Angewandte Chemie International Edition. 2016;**55**(31):8966-8969

[17] Wei D, Sang R, Sponholz P, Junge H, Beller M. Reversible hydrogenation of carbon dioxide to formic acid using a Mn-pincer complex in the presence of lysine. Nature Energy. 2022;7(5):438-447

[18] Akbayrak S, Özkar S. Ammonia borane as hydrogen storage materials. International Journal of Hydrogen Energy. 2018;**43**(40):18592-18606

[19] Huo J, Zhang K, Wei H, Fu L, Zhao C, He C, et al. A review on hydrogen production from ammonia borane: Experimental and theoretical studies. Chinese Chemical Letters. 2023:108280

[20] Sun D, Mazumder V, Metin Ö, Sun S. Methanolysis of ammonia borane by CoPd nanoparticles. ACS Catalysis. 2012; **2**(6):1290-1295

[21] Kumar DR, Prabu S, Chiang K-Y, Oh TH. Methanolysis of ammonia borane using binder-free hierarchical Co@Ni metal-organic framework nanocolumn arrays catalyst for hydrogen generation. International Journal of Energy Research. 2022;**46**(13):18134-18145

[22] Hao C, Wang S, Li M, Kang L, Ma X. Hydrogenation of CO_2 to formic acid on supported ruthenium catalysts. Catalysis Today. 2011;**160**(1):184-190

[23] Peters M, Köhler B, Kuckshinrichs W, Leitner W, Markewitz P, Müller TE. Chemical Technologies for Exploiting and Recycling Carbon Dioxide into the value chain. ChemSusChem. 2011;**4**(9): 1216-1240

[24] Markewitz P, Kuckshinrichs W, Leitner W, Linssen J, Zapp P, Bongartz R, et al. Worldwide innovations in the development of carbon capture technologies and the utilization of CO_2. Energy & Environmental Science. 2012;**5** (6):7281-7305

[25] Artz J, Müller TE, Thenert K, Kleinekorte J, Meys R, Sternberg A, et al. Sustainable conversion of carbon dioxide: An integrated review of catalysis and life cycle assessment. Chemical Reviews. 2018;**118**(2):434-504

[26] Langanke J, Wolf A, Hofmann J, Böhm K, Subhani MA, Müller TE, et al. Green Chemistry. 2014;**16**:1865-1870

[27] Zhang W, Lü X. Synthesis of Carboxylic Acids and Derivatives Using CO2 as Carboxylative Reagent. Chinese Journal of Catalysis. 2012;**33**(4–6): 745-756

[28] Álvarez A, Bansode A, Urakawa A, Bavykina AV, Wezendonk TA, Makkee M, et al. Challenges in the greener production of Formates/formic acid, methanol, and DME by heterogeneously catalyzed CO_2 hydrogenation processes. Chemical Reviews. 2017;**117**(14): 9804-9838

[29] Mariyaselvakumar M, Kadam GG, Mani M, Srinivasan K, Konwar LJ. Direct hydrogenation of CO_2-rich scrubbing solvents to formate/formic acid over heterogeneous Ru catalysts: A sustainable approach towards

continuous integrated CCU. Journal of CO2 Utilization. 2023;**67**:102326

[30] Moret S, Dyson PJ, Laurenczy G. Direct synthesis of formic acid from carbon dioxide by hydrogenation in acidic media. Nature Communications. 2014;**5**(1):4017

[31] Weilhard A, Dupont J, Sans V. 17. Carbon dioxide hydrogenation to formic acid. In: North M, Styring P, editors. Carbon Dioxide Utilisation. Transformations. 2. Berlin/Boston: De Gruyter; 2019. pp. 329-344

[32] Filonenko GA, van Putten R, Schulpen EN, Hensen EJM, Pidko EA. Highly efficient reversible hydrogenation of carbon dioxide to formates using a ruthenium PNP–pincer catalyst. ChemCatChem. 2014;**6**: 1526-1530

[33] Hafeez J, Bilal M, Rasool N, Hafeez U, Shah SAA, Imran S, et al. Synthesis of ruthenium complexes and their catalytic applications: A review. Arabian Journal of Chemistry. 2022;**15**(11) 104165:1-94

[34] Fernández-Alvarez FJ, Oro LA. Iridium-Catalyzed Homogeneous Hydrogenation and Hydrosilylation of Carbon Dioxide. In: Oro LA, Claver C, editors. Iridium Catalysts for Organic Reactions. Topics in Organometallic Chemistry. Vol. 69. Cham: Springer; 2020. pp. 303-324

[35] Cui M, Qian Q, Zhang J, Wang Y, Asare Bediako BB, Liu H, et al. Liquid fuel synthesis via CO_2 hydrogenation by coupling homogeneous and heterogeneous catalysis. Chem. 2021;**7** (3):726-737

[36] Prieto G. Carbon dioxide hydrogenation into higher hydrocarbons and oxygenates: Thermodynamic and kinetic bounds and Progress with heterogeneous and homogeneous catalysis. ChemSusChem. 2017;**10**(6): 1056-1070

[37] Melchionna M, Fornasiero P. Dual catalysis by homogeneous/ heterogeneous ruthenium species. Chem. 2021;**7**(4):834-835

[38] Srivastava V. Active heterogeneous Ru Nanocatalysts for CO_2 hydrogenation reaction. Catalysis Letters. 2016;**146**(12): 2630-2640

[39] Gunasekar GH, Padmanaban S, Park K, Jung K-D, Yoon S. An efficient and practical system for the synthesis of *N*, *N*-Dimethylformamide by CO_2 hydrogenation using a heterogeneous Ru catalyst: From batch to continuous flow. ChemSusChem. 2020;**13**(7):1735-1739

[40] Gautam P, Upadhyay PR, Srivastava V. Selective hydrogenation of CO_2 to formic acid over alumina-supported Ru nanoparticles with multifunctional ionic liquid. Catalysis Letters. 2019;**149**(6): 1464-1475

[41] Coeck R, Berden S, De Vos DE. Sustainable hydrogenation of aliphatic acyclic primary amides to primary amines with recyclable heterogeneous ruthenium–tungsten catalysts. Green Chemistry. 2019;**21**(19):5326-5335

[42] Li L, Shi J-l. A highly active and reusable heterogeneous ruthenium catalyst for olefin metathesis. Advanced Synthesis & Catalysis. 2005;**347**(14): 1745-1749

[43] Krause JO, Nuyken O, Wurst K, Buchmeiser MR. Synthesis and reactivity of homogeneous and heterogeneous ruthenium-based metathesis catalysts containing electron-withdrawing ligands. Chemistry – A European Journal. 2004;**10**(3):777-784

[44] Assmann J, Narkhede V, Breuer NA, Muhler M, Seitsonen AP, Knapp M, et al. Heterogeneous oxidation catalysis on ruthenium: Bridging the pressure and materials gaps and beyond. Journal of Physics: Condensed Matter. 2008;**20** (18):184017-184039

[45] Arends IWCE, Sheldon RA. Activities and stabilities of heterogeneous catalysts in selective liquid phase oxidations: Recent developments. Applied Catalysis A: General. 2001;**212**(1):175-187

[46] Muhammad S, Shukla PR, Tadé MO, Wang S. Heterogeneous activation of peroxymonosulphate by supported ruthenium catalysts for phenol degradation in water. Journal of Hazardous Materials. 2012;**215-216**: 183-190

[47] Ji H, Mizugaki T, Ebitani K, Kaneda K. Highly efficient oxidation of alcohols to carbonyl compounds in the presence of molecular oxygen using a novel heterogeneous ruthenium catalyst. Tetrahedron Letters. 2002;**43**(40): 7179-7183

[48] Gorbanev YY, Kegnæs S, Hanning CW, Hansen TW, Riisager A. Acetic acid formation by selective aerobic oxidation of aqueous ethanol over heterogeneous ruthenium catalysts. ACS Catalysis. 2012;**2**(4):604-612

[49] Yamaguchi K, Mizuno N. Supported ruthenium catalyst for the heterogeneous oxidation of alcohols with molecular oxygen. Angewandte Chemie International Edition. 2002;**41**(23): 4538-4542

[50] Melián-Rodríguez M, Saravanamurugan S, Kegnæs S, Riisager A. Aerobic oxidation of Veratryl alcohol to Veratraldehyde with heterogeneous ruthenium catalysts. Topics in Catalysis. 2015;**58**(14):1036-1042

[51] Mägerlein W, Dreisbach C, Hugl H, Tse MK, Klawonn M, Bhor S, et al. Homogeneous and heterogeneous ruthenium catalysts in the synthesis of fine chemicals. Catalysis Today. 2007; **121**(1):140-150

[52] Warratz S, Burns DJ, Zhu C, Korvorapun K, Rogge T, Scholz J, et al. Meta-C—H Bromination on purine bases by heterogeneous ruthenium catalysis. Angewandte Chemie International Edition. 2017;**56**(6):1557-1560

[53] Epp K, Luz I, Heinz WR, Rapeyko A, Llabrés i Xamena FX, Fischer RA. Defect-engineered ruthenium MOFs as versatile heterogeneous hydrogenation catalysts. ChemCatChem. 2020;**12**(6): 1720-1725

[54] Widegren JA, Bennett MA, Finke RG. Is it homogeneous or heterogeneous catalysis? Identification of bulk ruthenium metal as the true catalyst in benzene hydrogenations starting with the monometallic precursor, Ru(II)(η6-C6Me6)(OAc)2, plus kinetic characterization of the heterogeneous nucleation, then autocatalytic surface-growth mechanism of metal film formation. Journal of the American Chemical Society. 2003;**125**(34): 10301-10310

[55] Kraynov A, Müller TE. Concepts for the stabilization of metal nanoparticles in ionic liquids. Applications of Ionic Liquids in Science and Technology. 2011; **9**:235-260

[56] Gorbanev YY, Kegnæs S, Riisager A. Effect of support in heterogeneous ruthenium catalysts used for the selective aerobic oxidation of HMF in water. Topics in Catalysis. 2011;**54**(16): 1318

[57] Su F, Lv L, Lee FY, Liu T, Cooper AI, Zhao XS. Thermally reduced ruthenium nanoparticles as a highly active heterogeneous catalyst for hydrogenation of Monoaromatics. Journal of the American Chemical Society. 2007;**129**(46):14213-14223

[58] Kotani M, Koike T, Yamaguchi K, Mizuno N. Ruthenium hydroxide on magnetite as a magnetically separable heterogeneous catalyst for liquid-phase oxidation and reduction. Green Chemistry. 2006;**8**:735-741

[59] Jaleel A, Haider A, Nguyen CV, Lee KR, Choung S, Han JW, et al. Structural effect of nitrogen/carbon on the stability of anchored Ru catalysts for CO_2 hydrogenation to formate. Chemical Engineering Journal. 2022;**433**:133571

[60] Baig RBN, Nadagouda MN, Varma RS. Ruthenium on chitosan: A recyclable heterogeneous catalyst for aqueous hydration of nitriles to amides. Green Chemistry. 2014;**16**(4):2122-2127

[61] Anjali K, Christopher J, Sakthivel A. Ruthenium-based macromolecules as potential catalysts in homogeneous and heterogeneous phases for the utilization of carbon dioxide. ACS Omega. 2019;**4**(8):13454-13464

[62] Choi E, Lee C, Na Y, Chang S. [RuCl2(p-cymene)]2 on carbon: An efficient, selective, reusable, and environmentally versatile heterogeneous catalyst. Organic Letters. 2002;**4**(14):2369-2371

[63] Adam F, Balakrishnan S, Wong P-L. Rice husk ash silica as a support material for ruthenium based heterogenous catalyst. Journal of Physical Science. 2006;**17**(2):1-13

[64] Gan W, Dyson PJ, Laurenczy G. Heterogeneous silica-supported ruthenium phosphine catalysts for selective formic acid decomposition. ChemCatChem. 2013;**5**(10):3124-3130

[65] Adams RD, Trufan E. Ruthenium–tin cluster complexes and their applications as bimetallic nanoscale heterogeneous hydrogenation catalysts. Philosophical Transactions of the Royal Society A: Mathematical, Physical and Engineering Sciences. 2010;**368**(1915):1473-1493

[66] Sreenavya A, Ahammed S, Ramachandran A, Ganesh V, Sakthivel A. Nickel–ruthenium bimetallic species on Hydrotalcite support: A potential hydrogenation catalyst. Catalysis Letters. 2022;**152**(3):848-862

[67] Tomkins P, Gebauer-Henke E, Müller TE. Molecular assembly line: Stepwise hydrogenation of multifunctional substrates over catalyst mixtures. ChemCatChem. 2016;**8**(3):546-550

[68] Müller TE. Hydrogenation and Hydrogenolysis with ruthenium catalysts and application to biomass conversion. In: Ishida H, editor. Ruthenium: An Element Loved by Researchers. 2021;**6**:1-32